System Engineering and Automation

Javier Fernández de Cañete, Cipriano Galindo,
and Inmaculada García Moral

System Engineering and Automation

An Interactive Educational Approach

Springer

Authors

Javier Fernández de Cañete
University of Málaga
Engineering System &
Automation Dept.
C/Dr. Ortiz Ramos, s/n
29071 Málaga
Spain
E-mail: canete@isa.uma.es

Inmaculada García Moral
University of Málaga
Engineering System &
Automation Dept.
C/Dr. Ortiz Ramos, s/n
29071 Málaga
Spain
E-mail: gmoral@isa.uma.es

Cipriano Galindo
University of Málaga
Engineering System &
Automation Dept.
C/Dr. Ortiz Ramos, s/n
29071 Málaga
Spain
E-mail: cipriano@ctima.uma.es

ISBN 978-3-642-20229-2 e-ISBN 978-3-642-20230-8

DOI 10.1007/978-3-642-20230-8

Library of Congress Control Number: 2011925363

Typeset & Cover Design: Scientific Publishing Services Pvt. Ltd., Chennai, India.

Printed on acid-free paper

9 8 7 6 5 4 3 2 1

springer.com

To my wife and to my children for their understanding and to God for his help. J.F.C.

To my beloved wife Ana Belén, for her love and patience. C.G.

To my husband and my children for their patience and help. I.G.

Preface

Automatic Control is essentially a branch of engineering that addresses the problem of endowing systems with an autonomous capacity of interacting with the environment, while keeping constant its main variables values. It is applied to a wide variety of engineering fields, enabling an interdisciplinary approach between related fields.

This text is designed to provide insight and enhanced appreciation of analysis, modeling and control of dynamic systems. The reader is assumed to be familiar with calculus, physics and some programming skills. The perspective of the text is based on the system behavior in the time domain both, linear and non linear, and it supposes a focused presentation which develops the reader's ability to interpret physical significance of mathematical results in system analysis. This text carefully prepares the reader for more advanced treatment for subsequent knowledge in the automatic control field.

Learning objectives are performance-oriented, using for this purpose interactive MATLAB, SIMULINK and SIMSCAPE software tools, presenting also realistic problems in order to analyze, design and develop automatic control systems. Learning with computing tools can aid theory and help students to think, analyze and reason in meaningful ways. The text also is complemented with classroom slides and MATLAB, SIMULINK and SIMSCAPE exercise files to aid students to focus on the fundamental concepts treated.

The pedagogical approach hereby used distinguishes form other systems textbooks, because fundamental concepts are introduced together with adequate software tools in order to gain insight in many of the industrial engineering disciplines covered by the text.

This text comes out of more than 10 years of experience teaching System Engineering and Automation Control in the Computer Science Faculty, and is intended to serve as material base to teach a 6 ECTS credit subject following the EHEA (European Higher Education Area) guidelines.

The book starts with Chapter 1, where basic definitions and main concepts of System Engineering are introduced. This chapter gives a classification of different *types of systems*, focusing on dynamic, deterministic and continuous systems, and presents examples of both, *internal and external representations* of mechanical and electric systems. Although the book will address *SISO systems* (single input - single output) definitions are generalized to also cope with *MIMO systems* (multiple input - multiple output).

Chapter 2 introduces the concept of the *model of a system*, as a representation of a real or intended system. It details system modeling techniques to obtain the *differential equations of systems* of very diverse nature: electrical, mechanical, hydraulics, thermal, and hybrid ones. Along this chapter, modeling examples of each type of system are illustrated and implemented using SIMSCAPE and SIMULINK.

Once it is explained how models of systems can be obtained, Chapter 3, introduces the reader into the basics of the System and Automation Control Engineering through the concept of the *transfer function*. The transfer function converts differential equations into algebraic expressions which largely facilitate the study of the behavior of linear and time invariant systems. This chapter presents the *Laplace transformation* and its properties through a variety of examples. It ends illustrating *block diagrams* and the rules for their simplification, permitting the reduction of complex systems to a unique transfer function.

Chapter 4 copes with the *response analysis* of typical first order, second order and n-order systems. The temporal characteristics involved in systems' responses are studied and related to the roots of their characteristic equations (poles). Such relation is further analyzed by means of the *root locus* technique. The root locus depicts the place where the poles can be located as long as a system parameter varies. The rules for its construction are detailed while exercises are presented using the *rltool* graphical interface of MATLAB. This chapter also addresses the *stability* as well as the *steady-state error* concepts which become indispensable in any control application.

An introduction to *control systems* is given in Chapter 5, motivating the utilization of controllers, i.e. particular systems that modify the behavior of others. It focuses on their design and study by using the techniques and concepts covered in the previous chapters. Special attention is paid to the most spread controller, the *PID controller*, which is being using in most of the modern systems since the 30's. A detailed analysis of its three components is presented through practical exercises on MATLAB and SIMULINK.

Chapter 6 introduces the *system simulation*, its objectives and benefits. From the modeling of a given system (real or intended), to the understanding of its internal characteristics, the last natural step is the simulation of its behavior against different input signals. Computer simulation brings us the possibility of testing avoiding the need of acting on the real system, with the obvious advantages, e.g. a wrong decision taken on simulation will not cause any damage to the real system. In this chapter simulation of linear and non-linear systems is approached through the SIMULINK and SIMSCAPE computer tools. The former is well-known a widely used tool for system simulation although it exhibits the disadvantage of requiring the dynamic equations involved in the system which normally limits the complexity of the systems at hand. In contrast, SIMSCAPE permits the designer, as well as non-expert users, to model and simulate complex systems through a library of blocks that represent real components and their implicit dynamics.

The remaining is divided in four appendices with user manuals and guidelines for the software tools employed along the examples and exercises of this book.

Javier Fernández de Cañete
Cipriano Galindo
Inmaculada García Moral

Contents

Chapter 1
Introduction to Systems

1.1 System Concept

System Engineering refers to a discipline that aims to model, analyze, describe, and simulate physical *systems*. The term *system*, which comes from the Greek word *systema*, can be defined as a collection of inter-related elements pursuing a particular objective.

Colloquially, the concept of system is widely used in a variety of situations including, in some cases, the word *system* in the name. For example, the *solar system*, considered as a set of planets and satellites that orbit around the Sun; broadly speaking, the gravity force is the principal link that keeps their relative positions, allowing us to foresee, i.e. simulate, the planets' positions. Another commonly-known example is the *nervous system* composed of neurons, dendrites, and axons with the aim of transmitting electrical signals from our brain to the muscles. Also, the *political system* of a country fits the given definition of system as it is composed of a number of parties and political and administrative divisions with the objective of governing a nation.

Certainly, the concept of system is widespread and applicable to fields of very different nature. However, System Engineering deals with the study of systems which can be modeled by physical laws, (it is not in charge of cases like the third example exposed before), by providing a set of mathematical methodologies.

When studying systems, under the System Engineering perspective, one of the first steps to be given is to obtain a mathematical model, that is, a set of equations that capture the dynamics of the system at hand. To do that, physical laws are applied to obtain the correspondent differential equations.

For example, the shock absorber system of a vehicle depicted in fig. 1.1 can be modeled as

$$M\frac{d^2x}{dt^2} + B\frac{dx}{dt} + Kx(t) = F(t) \tag{1.1}$$

where M is the mass of the vehicle, x its position, B and K the constants of the mechanical elements involved, and $F(t)$ is the gravitational force. Imagine that we would like to check whether the current system is suitable for a weightier vehicle.

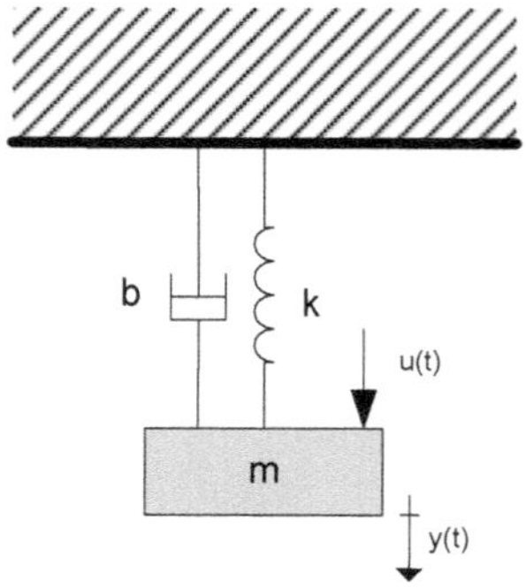

Fig. 1.1 Diagram of a simplified shock absorber mechanical system.

We could strip the shock absorber of the vehicle and install it on the new one and directly check out the results, or we could mathematically analyze the system, solve the equation for a different value of *M*, i.e. the mass of the new vehicle, and test the behavior of the system in simulation.

System Engineering offers to designers the possibility of easily managing mathematical models in order to:

1. Study the transient response of the system against different input signals. The transient response is related to the behavior of a given system after receiving a stimulus which gives us information about how quick the system responds, and how smooth is such a response. Following with the example of the shock absorber, we would be interested on a system that rapidly absorbs potholes without causing vibrations on the vehicle.
2. Study the stability and steady response of systems, that is, the convergence of the system output to a final value fulfilling the given requirements. For instance, the output of a heating system, e.g. the temperature of a room, should be stabilized exactly on the reference input by the user.
3. Modify given systems to behave in the desired manner, by adding particular systems, called *controllers*, that guide the system outputs.

1.2 Systems Classification

As commented, systems may be of a very diverse nature and thus classing them is needed for a better understanding and study. The following classification, relating to different system particularities is normally applied.

1.2.1 Static/Dynamic

According to the dependence of the current outputs of a system with respect to previous ones, systems can be classed as *static systems*, those systems without memory, i.e. the system output only depends on the current input, and *dynamic systems*, that is, if past outputs influence on the current one. For instance, an automatic door opening mechanism is a static system whose behavior does not depend on the past, i.e. previous arriving of people. In contrast, positioning system of a satellite is an example of a dynamic system, since the current position depends on the previous one, as well as on the previous values of speed and acceleration.

1.2.2 Stochastic/Deterministic

A system is said to be *stochastic* when it exhibits a certain random, unforeseeable behavior, and thus, it may output different responses for the same stimuli. In contrast, a system is said to be *deterministic* when no randomness is involved in its dynamics and therefore it always produces the same output for the same input. The systems given in the previous paragraph, automatic door and satellite positioning, are both deterministic system, while the output of a dice after throwing it is a stochastic example.

1.2.3 Continuous/Discrete/Event Discrete

Depending on the nature of the magnitudes involved in a system, it can be classed as *continuous*, *discrete*, or *event discrete* system.

A *continuous system* only involves continuous magnitudes, that is, variables that always has a certain value, i.e. $x=x(t), -\infty < t < \infty$. For example, if we think on magnitudes like temperature, position, speed, pressure, etc., all of them are continuous and could be plotted, having a graphical representation like the one shown in fig. 1.2. Actually, every physical magnitude in the nature is considered as continuous.

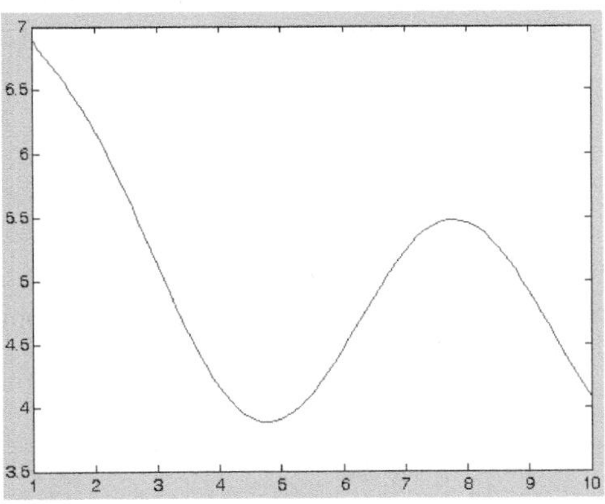

Fig. 1.2 Representation of a continuous magnitude.

A *discrete system*, in contrast, involves magnitudes, that only have a value for discrete times, $x=x(k)$, $k=t_0,t_1,t_2, \ldots,t_T$. For instance, fig. 1.3 depicts a discrete magnitude that only has a value every 0.5 seconds. Technically speaking, there are no discrete magnitudes, but they are continuous variables measured at discrete time intervals. For example, fig.1.3 actually represents the same information that the variable plotted in fig. 1.2, but sampling it regularly.

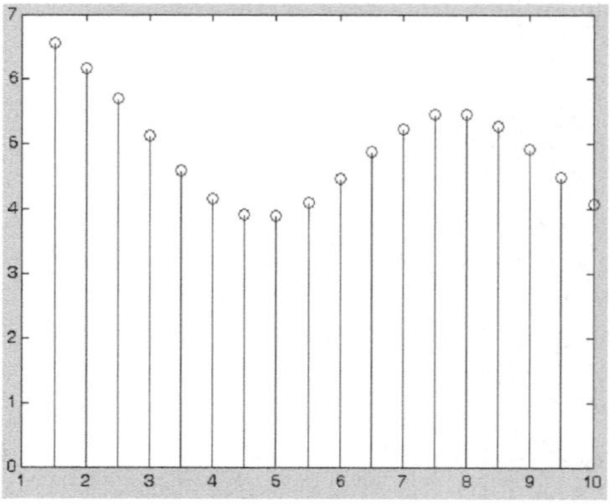

Fig. 1.3 Representation of a discrete magnitude. They normally come from samples of a continuous magnitude.

Finally, *event discrete* systems are a special case of discrete systems where the discrete magnitude can only take a finite of possible values (see fig. 1.4). For example the number of clients in a queue is an event discrete magnitude since its value will be $x_1, x_2, \ldots x_n$, $x \in N$.

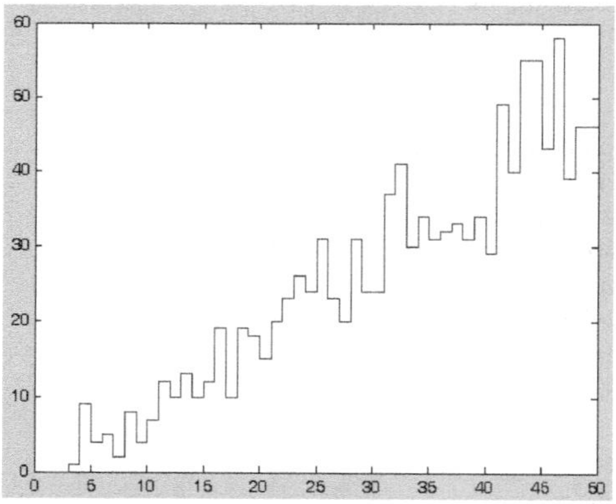

Fig. 1.4 Representation of an event discrete magnitude

1.2.4 *Closed/Open Loop*

According to the particular arrangement of the system, they can be classed as *open-loop systems* and *closed-loop systems*. This distinction is useful for control systems, that is, those which are aimed to produce an output that tends to the reference input. For instance, a heating device is a control system that gives a reference input, i.e. the desired temperature for a room, tries to produce an output that tends to such an input.

In open-loop control systems, the reference is input to the system, precisely to a subsystem called controller, and no feedback is considered (see fig. 1.5).

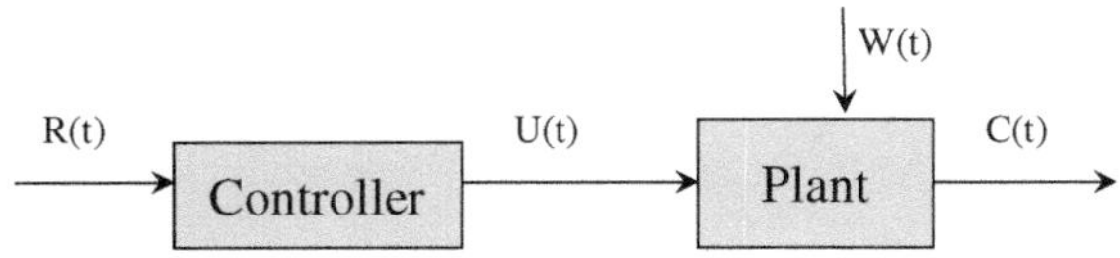

Fig. 1.5 Open-loop control configuration. The controller acts on the plant, i.e. the system to be controlled, producing a certain output *C(t)* which should tend to the reference. External disturbing signals, *W(t)* can negatively affect the plant.

In closed-loop control systems, as the one shown in fig. 1.6, a feedback line is considered permitting the system to measure the output and compare it with respect to the reference. The difference, called *error*, is the new input to the controller, adjusting thus the actuation on the system.

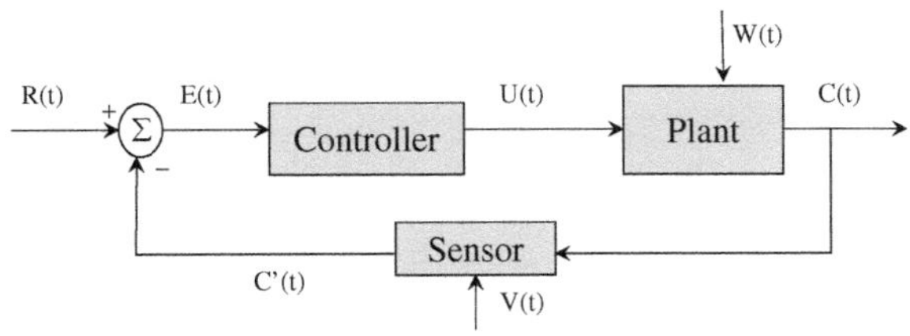

Fig. 1.6 Closed-loop control configuration. A feedback line is added, continuously comparing the output to the reference. Now, external disturbing signals on the plant, and also on the sensor will be considered by the controller, reducing their effect on the system output.

1.3 Systems Description

A system can be considered as a process that responds to a set of input signals or excitations $\bar{u}(t) = \{u_1(t), u_2(t), \dots, u_m(t)\}$ to produce a corresponding set of output signals or responses $\bar{y}(t) = \{y_1(t), y_2(t), \dots, y_n(t)\}$ when continuous systems are considered (see fig. 1.7).

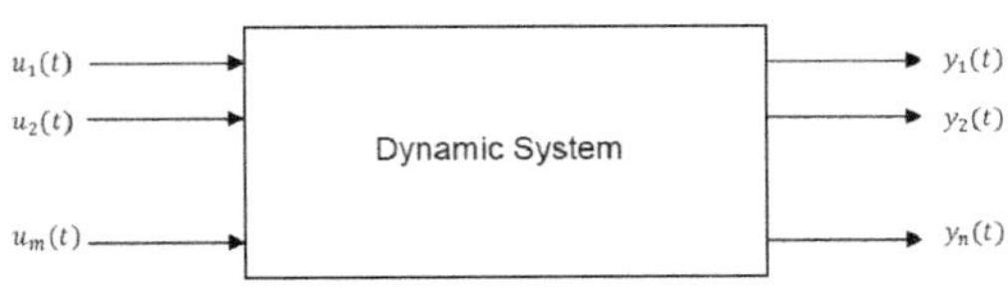

Fig. 1.7 Continuous multi-input, multi-output dynamic system.

Example 1.1

Define the set of input and output variables corresponding to six degrees of freedom manipulator as is described in fig. 1.8.

Solution

The set of input variables is given by the input signals to the robot drivers, so that, the input set is given by $\bar{\mathrm{u}}(\mathrm{t}) = \{\mathrm{v}_1(\mathrm{t}), \mathrm{v}_2(\mathrm{t}), \dots, \mathrm{v}_6(\mathrm{t})\}$ where $\mathrm{v}_\mathrm{i}(\mathrm{t}), \mathrm{i} = 1, \dots, 6$ corresponds to the input voltage, corresponding to each robot joint motor.

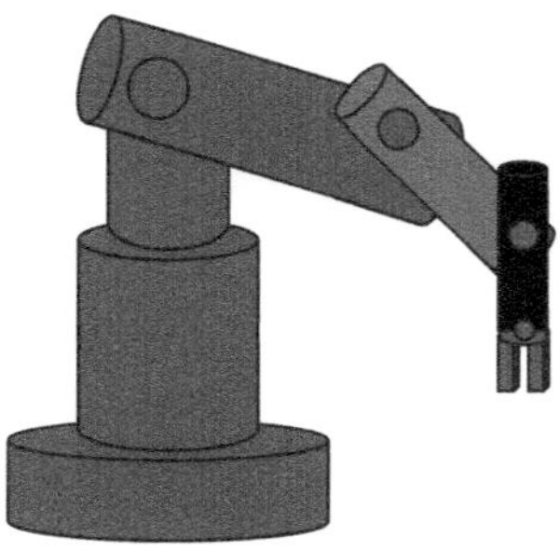

Fig. 1.8 Scheme of a manipulator of six degrees of freedom

The set of output variables is given by the output signals to the robot sensors, so that, the input set is given by $\bar{y}(t) = \{\theta_1(t), \theta_2(t), \dots, \theta_6(t)\}$ where $\theta_i(t), i = 1, \dots, 6$ corresponds to the angle corresponding to each robot joint.

In order to describe the system behavior we use a simplified representation of the dynamic system under analysis, which is called the system model.

There are two types of descriptions of a system, termed external and internal description, depending on the direct or indirect relationship between input and output signal of the system.

1.3.1 External Description

The external description is based on a black-box type structure (see fig. 1.9) relating the system output $\bar{y}(t) = \{y_1(t), y_2(t), \dots, y_n(t)\}$ and its derivatives with the system input $\bar{u}(t) = \{u_1(t), u_2(t), \dots, u_m(t)\}$ and its derivatives through a set of s independent input-output differential equations given by

$$
\begin{gathered}
f_1\left(y_1^{n)}, \dots, \dot{y}_1, y_1, \dots, y_s^{n)}, \dots, \dot{y}_s, y_s, u_1^{m)}, \dots, \dot{u}_1, u_1, \dots, u_r^{m)}, \dots, \dot{u}_r, u_r, t\right) = 0 \\
f_2\left(y_1^{n)}, \dots, \dot{y}_1, y_1, \dots, y_s^{n)}, \dots, \dot{y}_s, y_s, u_1^{m)}, \dots, \dot{u}_1, u_1, \dots, u_r^{m)}, \dots, \dot{u}_r, u_r, t\right) = 0 \\
\vdots \\
f_s\left(y_1^{n)}, \dots, \dot{y}_1, y_1, \dots, y_s^{n)}, \dots, \dot{y}_s, y_s, u_1^{m)}, \dots, \dot{u}_1, u_1, \dots, u_r^{m)}, \dots, \dot{u}_r, u_r, t\right) = 0
\end{gathered}
\tag{1.2}
$$

where $m \leq n$ for causal behavior, with $f_i, i = 1, \ldots, s$ non-linear in general. The parameter n is denoted as the system order, and plays a fundamental role for the system behavior.

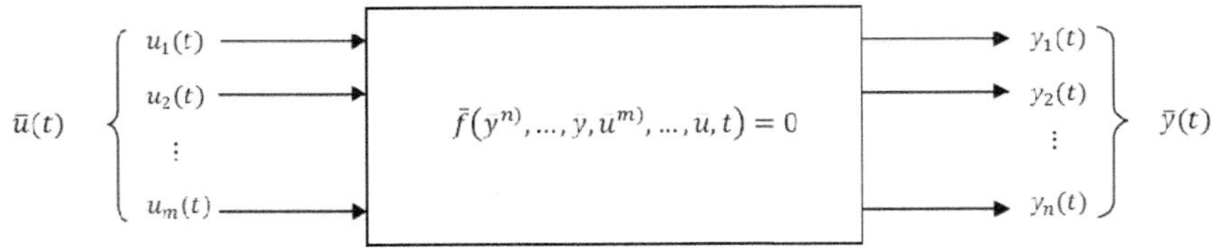

Fig. 1.9 External description of a continuous system.

If the system is single-input and single-output, the external description is given by

$$f\left(y^{n)}, \ldots, \dot{y}, y, u^{m)} \ldots, \dot{u}, u, t\right) = 0 \tag{1.3}$$

where $m \leq n$ for causal behavior, with f being non-linear in general.

The external description does not provide information on its internal structure. Moreover, the external description of different physical systems may be the same, which is the case of analog systems. The analogy is based on an underlying similarity between the forces that govern the behavior of both systems respectively.

Example 1.2

Demonstrate the relationship of analogy between the mechanical and electrical systems of fig. 1.10 based on differential equations that define its external description.

$$M\frac{d^2x}{dt^2} + D\frac{dx}{dt} + Kx(t) = F(t)$$

$$L\frac{d^2q}{dt^2} + R\frac{dq}{dt} + \frac{1}{C}q(t) = E(t)$$

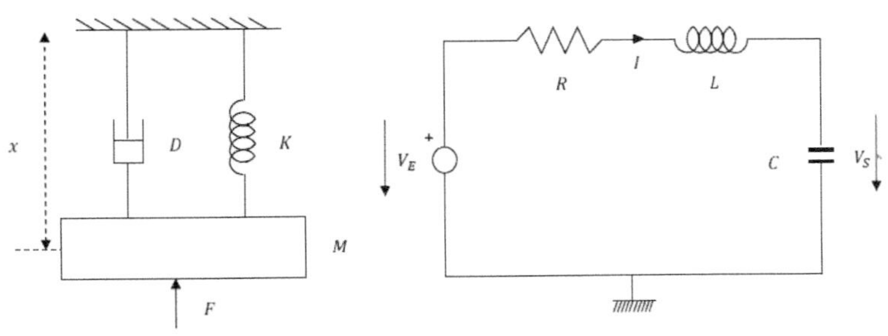

Fig. 1.10 Schematic of analogous mechanical and electric systems

Solution

Comparing both differential equations, it is clear that both systems are analogous. It is also possible to establish a table of equivalences between the variables and parameters of both analog systems (table 1.1).

Table 1.1 Equivalences by analogy between mechanical and electrical systems

Mechanical System		Electrical System	
displacement	$x(t)$	$q(t)$	charge
velocity	$v(t) = \frac{dx}{dt}$	$i(t) = \frac{dq}{dt}$	current
force	$F(t)$	$E(t)$	voltage
mass	M	L	inductance
damping	D	R	resistance
stiffness	K	$\frac{1}{C}$	capacitance^{-1}

1.3.2 *Internal Description*

The internal description is based on the indirect relationship between the input $\bar{u}(t)$ and the output $\bar{y}(t)$ of the system through the system state (fig. 1.11).

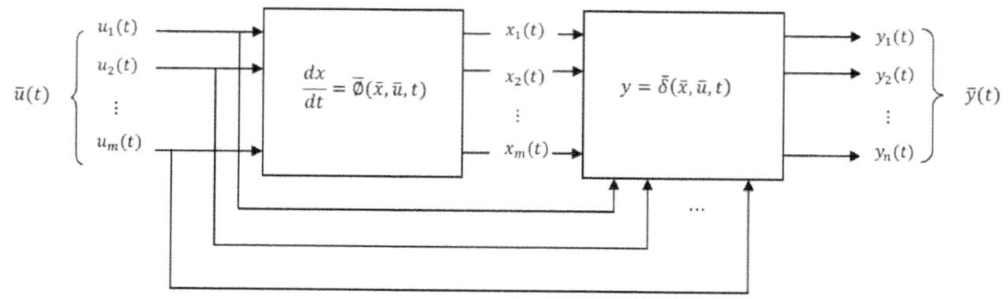

Fig. 1.11 Internal description of a continuous system.

The state of a dynamic system is defined by the minimum set of variables $\bar{x}(t) = \{x_1(t), x_2(t), \ldots, x_n(t)\}$ such that the knowledge of the value of $\bar{x}(t_0)$ together with the knowledge of the input $\bar{u}(t), t \geq t_0$, completely determines the system behavior for $t \geq t_0$.

The internal description is given in general by the state transition and the state output equations. The state transition equations can be described by a set of n independent differential equations in the following form

$$\begin{aligned} \dot{x}_1 &= \emptyset_1(x_1, x_2, \ldots, x_n, u_1, u_2, \ldots, u_m, t) \\ \dot{x}_2 &= \emptyset_2(x_1, x_2, \ldots, x_n, u_1, u_2, \ldots, u_m, t) \\ &\vdots \\ \dot{x}_n &= \emptyset_n(x_1, x_2, \ldots, x_n, u_1, u_2, \ldots, u_m, t) \end{aligned} \tag{1.4}$$

with $\emptyset_i, i = 1, \ldots, n$ non-linear in general. The parameter n is denoted as the system order, and plays a fundamental role for the system behavior.

In the same way, the state output equations can be described by a set of s independent differential equations in the following form

$$\begin{aligned} y_1 &= \delta_1(x_1, x_2, \ldots, x_n, u_1, u_2, \ldots, u_r, t) \\ y_2 &= \delta_2(x_1, x_2, \ldots, x_n, u_1, u_2, \ldots, u_r, t) \\ &\vdots \\ y_s &= \delta_s(x_1, x_2, \ldots, x_n, u_1, u_2, \ldots, u_r, t) \end{aligned} \tag{1.5}$$

also with $\delta_i, i = 1, \ldots, s$ non-linear in general.

Example 1.3

Determine the external and internal description corresponding to the electric circuit of fig. 1.12 with $V_E(t)$ and $V_s(t)$ considered as input u(t) and output y(t) of the system respectively and system state defined by $\bar{x}(t) = \{V_s(t), \dot{V}_s(t)\}$.

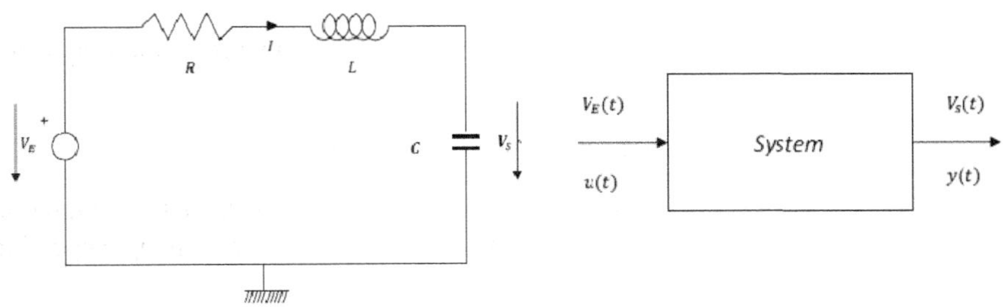

Fig. 1.12 Electric circuit for example 1.3

Solution

Applying the Kirchoff law for the electric circuit we have

$$RI(t) + L\frac{dI}{dt} + V_s(t) = V_E(t)$$

and the capacitor charge is given by

$$V_S(t) = \frac{1}{C}\int I(\tau)d\tau$$

Therefore, in order to derive the external description we must find the relationship between $V_S(t)$ and its derivatives and $V_E(t)$ and its derivatives. Thus, solving for $I(t)$ we have

$$I(t) = C\frac{dV_S}{dt}$$

Substituting this equation into the first one we finally obtain the external description, that is,

$$LC\frac{d^2V_s}{dt^2} + RC\frac{dV_s}{dt} + V_s(t) = V_E(t)$$

which can be expressed alternately as

$$f\left(\frac{d^2V_s}{dt^2}, \frac{dV_s}{dt}, V_S, V_E\right) = LC\frac{d^2V_s}{dt^2} + RC\frac{dV_s}{dt} + V_s(t) - V_E(t) = 0$$

$$f(\ddot{y}, \dot{y}, y, u) = LC\ddot{y} + RCy \dot{+} y - u = 0$$

In order to derive the internal description, we have to select the state variables, as many as the order of the system is, in this case $n = 2$, which coincides with the maximum order in the derivative of the output $V_S(t)$. In this way

$$\bar{x}(t) = \{x_1(t), x_2(t)\} = \{V_s(t), \frac{dV_s}{dt}(t)\}$$

Then, by using the state variables assignment and solving for $\frac{dV_S}{dt}(t)$ we have

$$\frac{dx_1}{dt} = \frac{dV_s}{dt} = x_2 = \emptyset_1(x_1, x_2, u)$$

$$\frac{dx_2}{dt} = \frac{d^2V_s}{dt^2} = \frac{1}{LC}(V_E(t) - RCx_2 - x_1) = \emptyset_2(x_1, x_2, u)$$

together with

$$y = V_s = x_1 = \delta(x_1, x_2, u)$$

Chapter 2
System Modeling

System modeling is rather a difficult task, since it requires experience, practice, and intuition to be a good system modeler. The bases for the building of mathematical models are the fundamental physical laws, such as the laws of conservation of mass, energy, and momentum.

In this chapter, we deal with the concept of model and illustrate the process to be followed to develop a model of a dynamical system in general. Also, the dynamical equations in the form of differential equations that describe the system will be obtained, as well as the transfer functions when linear behavior is present.

Finally, we end by describing the modeling of continuous systems corresponding to the electrical, mechanical, hydraulic, thermal, and hybrid systems among others. Some examples have been included to help the reader to clarify the methodology used here, to derive the dynamical equations, and we have also implemented these examples by using both the SIMULINK and SIMSCAPE modeling environments.

2.1 System Model

In almost all areas of engineering, considerable efforts are directed toward acquiring information on various aspects of system performance, which is called system analysis.

Traditionally, the system analysis was carried out using the physical system subjected to test input signals, observing its corresponding response. Nevertheless, this is not always feasible, since either the extent to which the parameters of the physical system can be varied is very limited or there are imposed restrictions to the application of these input tests (i.e. system analysis in human physiology).

In order to overcome these problems, a simplified representation of the physical system under analysis is developed instead, what is called system model.

The system model may be either mental, when expressed through an abstract description of the relationships among the system variables, or formal model, when using graphs, diagrams, and mathematical equations derived from the mental model used to describe the system behavior. The formal model is more explicit than a mental model, and its implementation in the computer produces the computer model.

The building process followed to obtain a computer model is represented in fig. 2.1. Starting from the observation of the real physical system behavior, and using information through expert opinion and the literature, a formal model is derived, by identifying the elements of the model and their interrelationships.

After establishing the formal diagrams relating to the formal model, we calculate the dynamic equations of the model and its implementation is done on computer by using an appropriate language to process the set of dynamic equations. The computer model is used to predict the behavior of the real system, so that we proceed to model validation by comparing the observed and predicted behavior, and in case of discrepancy, adjustment is made for improvement of the formal model.

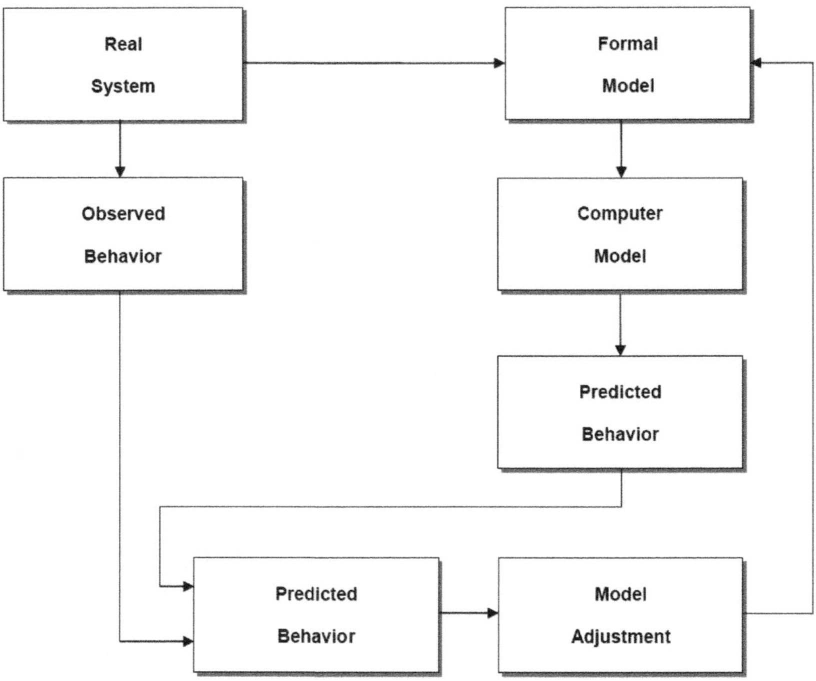

Fig. 2.1 Building process diagram for system modeling.

The process of building a system model is not linear, passing in successive stages by progressively improved models. In the successive stages is performed phasing more simplifying assumptions, so that the system model is moving ever closer to reality.

Thus, system models should be as simple as possible, and each model should be developed with a specific application in mind.

System models can be grouped according to several different criteria:

1. Linear/Nonlinear, depending on the application of the principle of superposition to system dynamic equations
2. Distributed/Lumped, depending on the variables dependency as function of spatial coordinates and time in dynamic equations.
3. Time-Varying/Stationary, depending on the variations of system model parameters in time, in dynamic equations.
4. Continuous/Discrete, depending on time variable definition over continuous o distinct values range.

The type of system models we are going to discuss are referred to linear, lumped, stationary, and continuous systems in the first part of the book, leaving the nonlinear and time-varying system behavior for the second part.

2.2 System Dynamic Equations

A methodology for developing system models described by dynamic equations is going to be developed and applied to electrical, mechanical, hydraulic, thermal, and hybrid systems.

These equations are going to be derived by decomposing the overall system in simple subsystems, so that the application of conservation laws (mass, momentum, energy ...) in each subsystem, together with the constitutive equations of each element will enable the obtaining of the dynamic equations. Finally, programming software as SIMULINK or SIMSCAPE will be used to derive the computer model.

For this purpose, the use of generalized variables will be considered, the variable flow (*f*) on the dynamic element and the variable effort (*e*) along the element, being the variable power (*P*) the product of both variables. Also we can define the variable displacement (*q*) as the integral of flow, the variable momentum (*p*) as the integral of effort and the variable energy (*E*) as the integral of power.

These relations between generalized variables can be represented as a thetraedron (fig. 2.2), and the application to electrical, mechanical, hydraulic, and thermal systems is straightforward by selecting adequately the corresponding generalized variables (fig. 2.3).

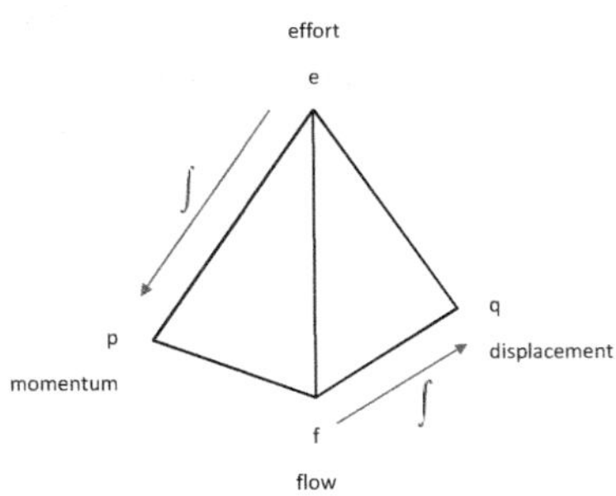

Fig. 2.2 Generalized variables used for system modeling and its relations.

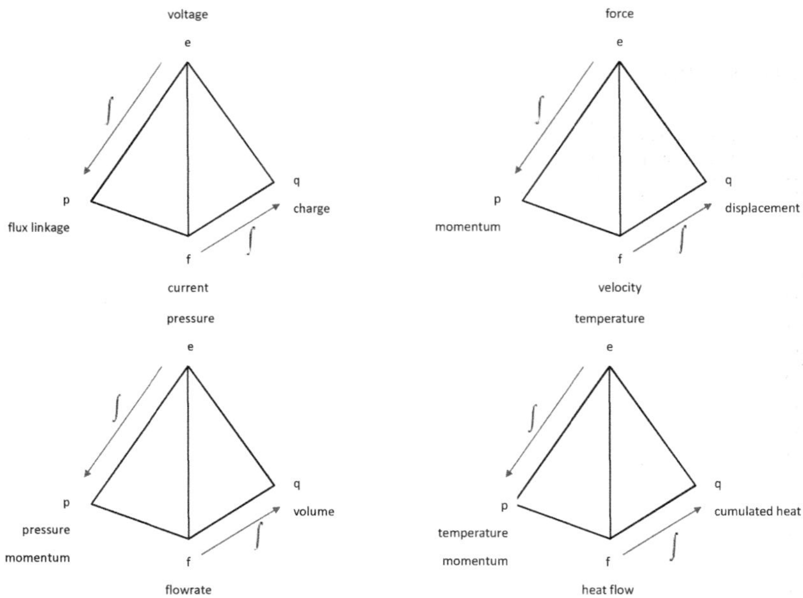

Fig. 2.3 Generalized variables relations for electrical, mechanical, hydraulic, and thermal systems

Similarly, the linkages between the generalized variables give rise to the concepts of resistance (*R*), capacitance (*C*), and inertia (*L*) characterizing the components of the system models (fig. 2.4).

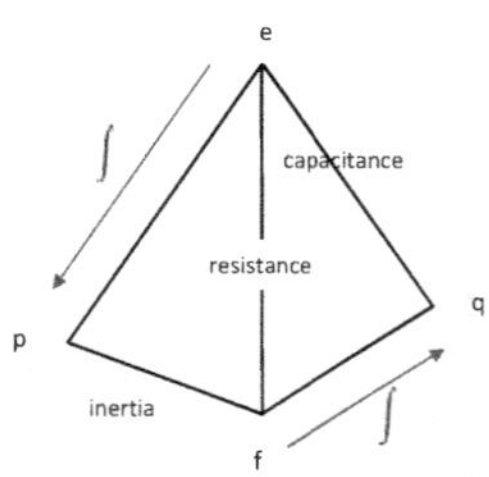

Fig. 2.4 Linkages between generalized variables for system models

The linkage between effort and flow variables defines the resistive element, generally a non-linear relationship, which in the linear case defines the resistance parameter *R* that characterizes the dynamic property of this element.

Table 2.1 defines the general nonlinear relationship both in direct and inverse form, the linear relationship, and the dimension of resistance for electrical systems, mechanical displacement and rotation systems, hydraulic systems and thermal systems.

Table 2.1 Resistive elements relations between generalized variables.

	General Relation	Linear Relation
Generalized Variables	$e = F_R(f)$ $f = {F_R}^{-1}(e)$	$e = Rf$ $f = \frac{e}{R} = Ge$
Electrical	$e = F_R(i)$ $i = {F_R}^{-1}(e)$	$e = Ri$ $i = \frac{e}{R} = Ge$
Mechanical Translation	$F = F_R(v)$ $v = {F_R}^{-1}(F)$	$F = Bv$ $v = \frac{F}{B}$
Mechanical Rotation	$\tau = F_R(\omega)$ $\omega = {F_R}^{-1}(\tau)$	$\tau = C\omega$ $\omega = \frac{\tau}{C}$
Hydraulic	$P = F_R(Q)$ $Q = {F_R}^{-1}(P)$	$P = RQ$ $Q = \frac{P}{R}$
Thermal	$T = F_R(h)$ $h = {F_R}^{-1}(T)$	$T = Rh$ $h = \frac{T}{R}$

Table 2.2 Capacitive elements relations between generalized variables.

	General Relation	Linear Relation
Generalized Variables	$q = F_C(e)$ $e = {F_C}^{-1}(q)$	$q = Ce$ $e = \frac{q}{C}$
Electrical	$q = F_C(e)$ $e = {F_C}^{-1}(q)$	$q = Ce$ $e = \frac{q}{C}$
Mechanical Translation	$x = F_R(F)$ $F = {F_R}^{-1}(x)$	$x = CF$ $F = kx$
Mechanical Rotation	$\theta = F_C(\tau)$ $\tau = {F_C}^{-1}(\theta)$	$\theta = C\tau$ $\tau = k\theta$
Hydraulic	$V = F_C(P)$ $P = {F_C}^{-1}(V)$	$V = CP$ $P = \frac{V}{C}$
Thermal	$Q = F_C(T)$ $T = {F_C}^{-1}(Q)$	$Q = CT$ $T = \frac{Q}{C}$

The linkage between effort and displacement variables defines the capacitive element, generally a non-linear relationship, which in the linear case defines the capacitance parameter C that characterizes the dynamic property of this element. table 2.2 defines the general nonlinear relationship both in direct and inverse form, the linear relationship and the dimension of capacitance for electrical systems, mechanical displacement and rotation systems, hydraulic systems, and thermal systems.

Finally, the linkage between effort and momentum variables defines, the inertia element, generally a non-linear relationship, which in the linear case, defines the inertia parameter I that characterizes the dynamic property of this element. table 2.3 defines the general nonlinear relationship both in direct and inverse form, the linear relationship and the dimension of capacitance for electrical systems, mechanical displacement and rotation systems, hydraulic systems, and thermal systems.

Table 2.3 Inertia element's relations between generalized variables.

	General Relation	**Linear Relation**
Generalized Variables	$p = F_I(f)$ $f = F_I^{-1}(p)$	$p = If$ $f = \frac{p}{I}$
Electrical	$\lambda = F_I(i)$ $i = F_I^{-1}(\lambda)$	$\lambda = Li$ $i = \frac{\lambda}{I}$
Mechanical Translation	$p = F_I(v)$ $v = F_I^{-1}(p)$	$p = mv$ $v = \frac{p}{m}$
Mechanical Rotation	$p_\tau = F_I(\omega)$ $\omega = F_I^{-1}(p_\tau)$	$p_\tau = J\omega$ $\omega = \frac{p_\tau}{J}$
Hydraulic	$p_P = F_I(Q)$ $Q = F_I^{-1}(p_P)$	$p_P = IQ$ $Q = \frac{p_P}{I}$
Thermal	$p_T = F_I(h)$ $h = F_I^{-1}(p_T)$	$p_T = Ih$ $h = \frac{p_\tau}{I}$

In order to obtain the dynamic equations corresponding to these physical systems, conservation laws are applied involving the elements above described. Basically, the values of the variables effort of interrelated elements that form a loop is zero and the sum of the flow variables at a node of interconnection is also zero (fig. 2.5).

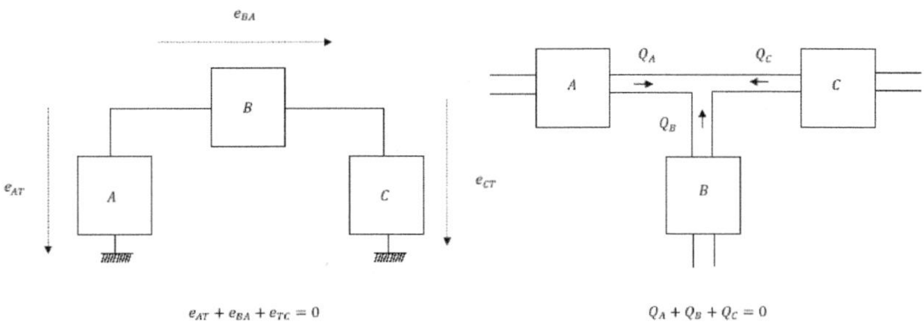

Fig. 2.5 Conservation laws used for dynamic equations of system models

2.3 Electrical Systems

In order to derive the mathematical model of electric systems, we will deal only with lumped systems made up of combinations of ideal electrical elements.

Table 2.4 Electrical elements and corresponding dynamic equations

Voltage Source	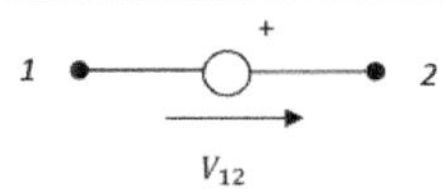	$v_{12} = v_{12}(t)$
Current Source	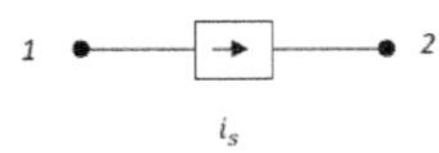	$i_S = i_S(t)$
Resistor	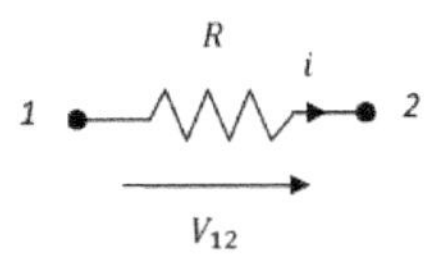	$v_{12} = R\,i$
Inductor	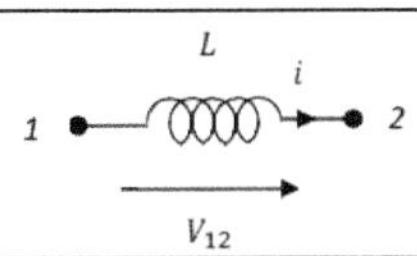	$v_{12} = L\frac{di}{dt}$
Capacitor	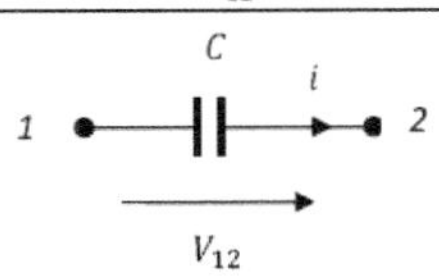	$\frac{dv_{12}}{dt} = \frac{1}{C}\,i$
Transformer	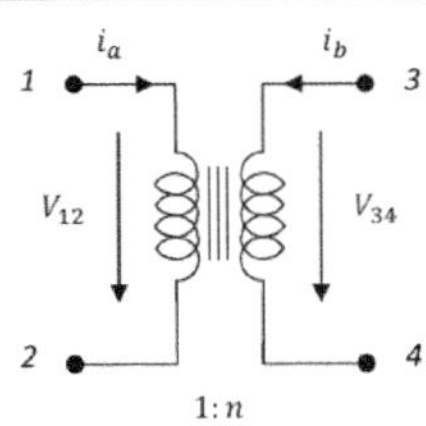	$v_{34} = n \cdot v_{12}$ $i_b = \frac{1}{n} i_a$

Most elements to be considered are two-terminal elements, characterized by the voltage and current, as effort and flow variables respectively.

In table 2.4 the main elements that appear at electrical circuits with its corresponding icons and dynamic equations are described.

The Kirchoff laws are employed to obtain a complete set of equations involving the electrical elements as they are connected to form the electrical system. The Kirchoff's voltage law settles that the sum of voltage that drops around a loop must be equal to zero, while the current law states that the sum of the currents at a node must be also zero (fig. 2.6).

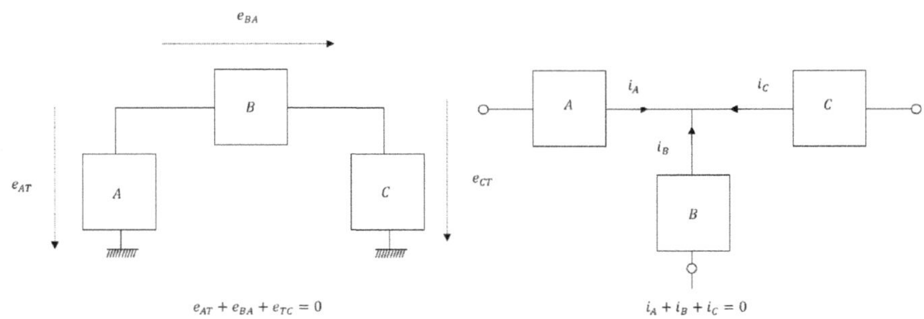

Fig. 2.6 Conservation laws used to obtain the dynamic equations for electrical systems.

Example 2.1

Derive the dynamic equations corresponding to the circuit shown in fig. 2.7 with $i_s(t)$ and $v_o(t)$ considered as input and output of the system respectively, also drawing the corresponding SIMULINK and SIMSCAPE modeling diagrams.

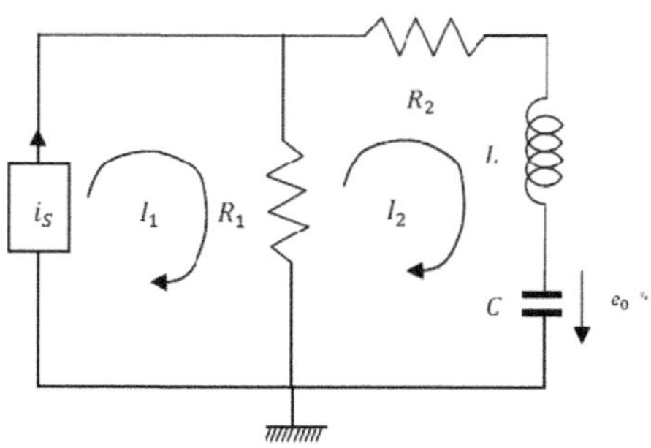

Fig. 2.7 Electric system for example 2.1

Solution

$$R_2 i_2 + L\frac{di_2}{dt} + v_o + R_1(i_2 - i_1) = 0$$

$$i_2 = C\frac{dv_0}{dt}$$

$$i_1 = i_s$$

Solving for i_2 and substituting i_1 into the first equation yields

$$LC\frac{d^2v_0}{dt^2} + (R_1 + R_2)C\frac{dv_0}{dt} + v_o = R_1 i_s$$

In fig. 2.8 and fig. 2.9 the corresponding modeling diagram developed by using both the SIMULINK and SIMSCAPE environments are shown.

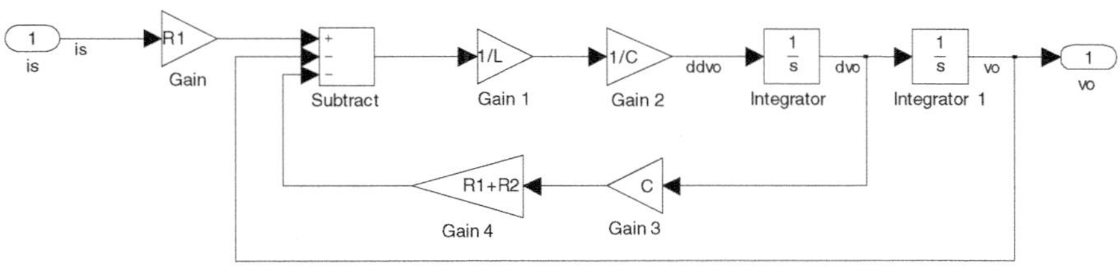

Fig. 2.8 SIMULINK diagram corresponding to example 2.1

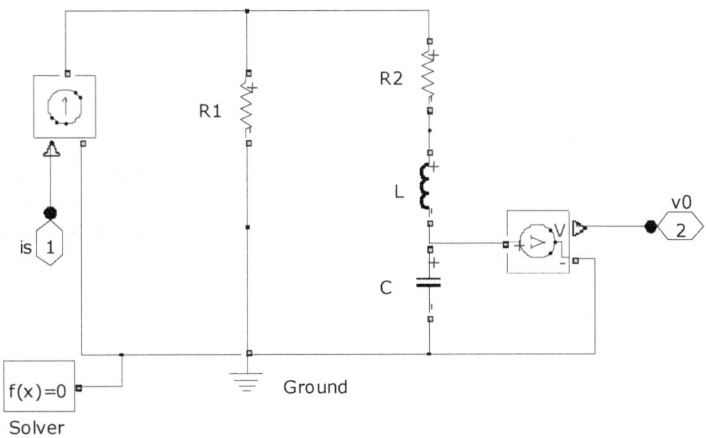

Fig. 2.9 SIMSCAPE diagram corresponding to example 2.1

Some important types of electric elements have more than two terminals, as is the case of the operational amplifier (OA), with two input terminals and one output terminal. This element's behavior can be approximated by an ideal equivalent circuit (fig. 2.10).

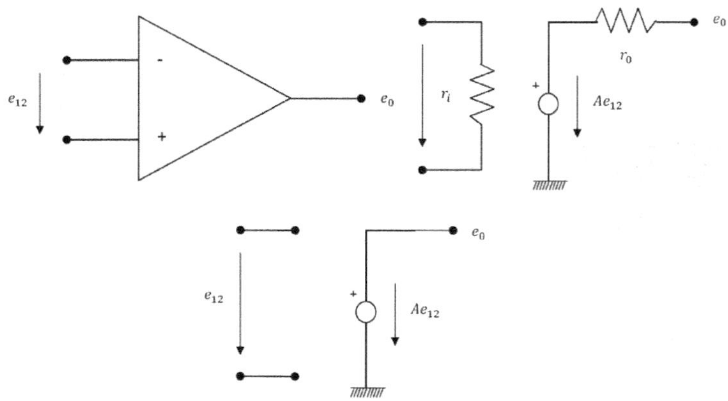

Fig. 2.10 Operational amplifier and equivalent circuit

Normally, the ideal operational amplifier is used in practice with feedback in order to stabilize it, from the start to the inverting terminal (fig. 2.11). Thus, assuming a zero voltage drop between inverting and non inverting terminals, we have

$$v_o(t) = -v_r(t) = -Ri_1(t)$$

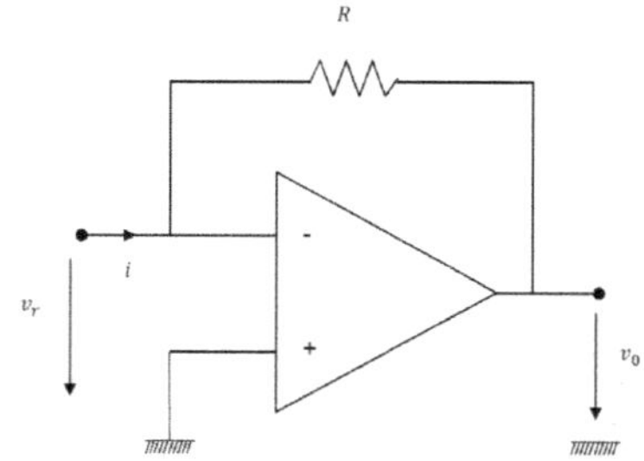

Fig. 2.11 Operational amplifier with feedback loop

Example 2.2

Derive the dynamic equations corresponding to the circuit with OA shown in fig. 2.12 with $e_i(t)$ and $e_o(t)$ considered as input and output of the system respectively, also drawing the corresponding SIMULINK and SIMSCAPE modeling diagrams.

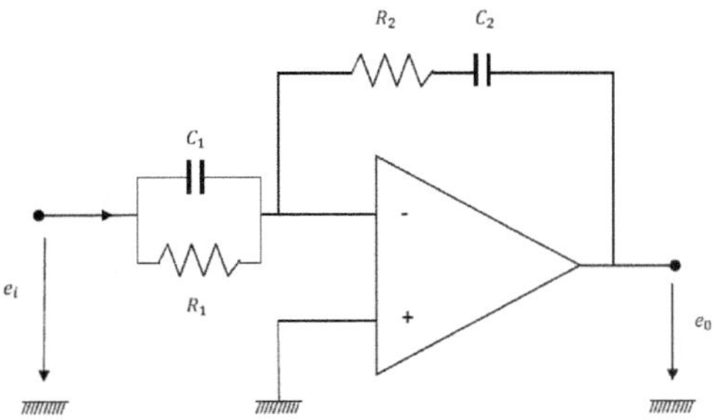

Fig. 2.12 Electric system for example 2.2

Solution

$$e_i = \frac{1}{C_1}\int i_{11}dt = R_1 i_{12}$$

$$i_1 = i_{11} + i_{12} = -i_2$$

$$e_0 = R_2 i_2 + \frac{1}{C_2}\int i_2\, dt$$

In fig. 2.13 and fig. 2.14 the corresponding modeling diagram developed by using both the SIMULINK and SIMSCAPE environments are shown.

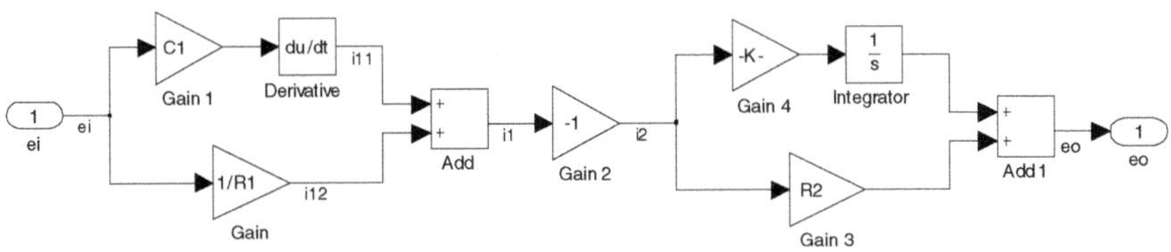

Fig. 2.13 SIMULINK diagram corresponding to example 2.2

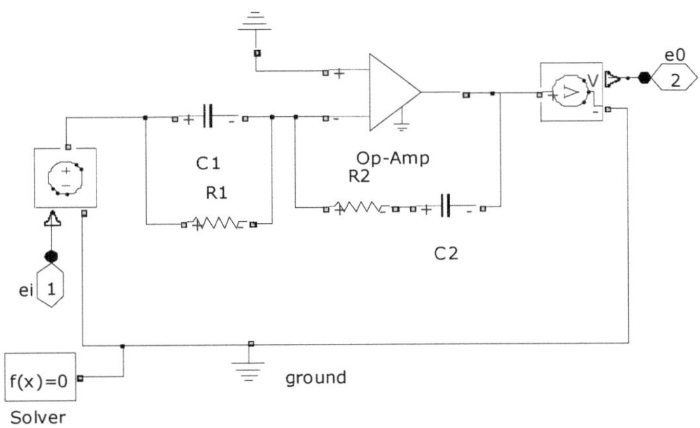

Fig. 2.14 SIMSCAPE diagram corresponding to example 2.2

2.4 Mechanical Systems

In order to obtain the dynamic equations governing the mechanical systems, we deal firstly with the translational mechanical systems, and secondly with the mechanical rotational systems. Again, we will deal only with lumped systems made up of combinations of ideal mechanical elements.

In table 2.5 the main elements that appear at translational mechanical systems with its corresponding icons and dynamic equations are described.

Table 2.5 Translational mechanical elements and corresponding dynamic equations

Mass	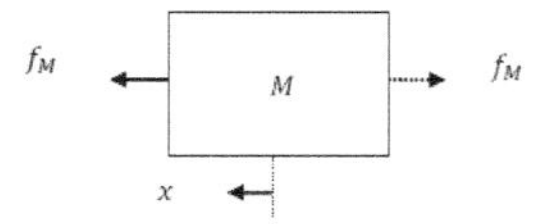	$f_M = M\frac{d^2x}{dt^2}$
Translational Spring	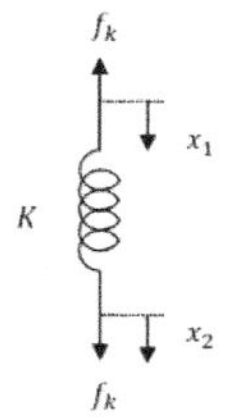	$f_K = K(x_1 - x_2)$
Translational Damper	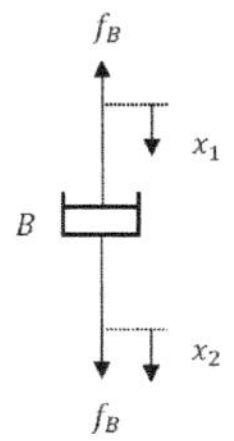	$f_B = B(\frac{dx_1}{dt} - \frac{dx_2}{dt})$
Lever	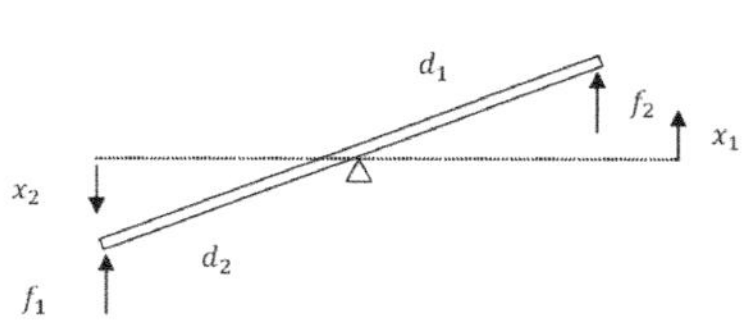	$f_2 = \frac{d_1}{d_2}f_1$ $x_2 = \frac{d_2}{d_1}x_1$

The D'Alambert law is applied to get the dynamic equations corresponding to the translational mechanical systems. This is just a restatement of the second law of Newton, described by

$$\sum_i f_{ext_i} = M\frac{d^2x_i}{dt^2} \quad (2.1)$$

where the summation includes all the external forces applied to the mass M. Then, by considering the term $f_M = M\frac{d^2x}{dt^2}$ as the inertial force, it results in the translational D'Alambert law given by

$$\sum_i f_{ext_i} = 0 \quad (2.2)$$

The dynamic equations are obtained by applying the D'Alembert law to each mass, using the free-body diagram showing all external forces and the inertial force with corresponding senses, once the positive sense has been previously defined.

Example 2.3

Derive the dynamic equations corresponding to the translational mechanical system shown in fig. 2.15, with fa(t) and x1(t) with x2(t) considered as input and output of the system respectively, also drawing the corresponding SIMULINK and SIMSCAPE modeling diagrams.

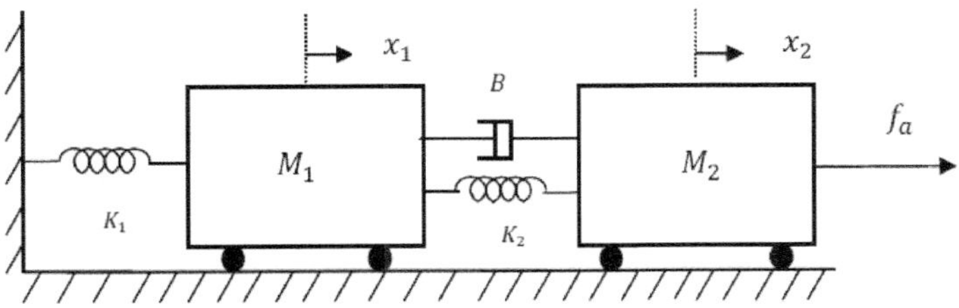

Fig. 2.15 Translational mechanical system for example 2.3

Solution

Firstly, we draw the free-body diagram for each mass (fig. 2.16)

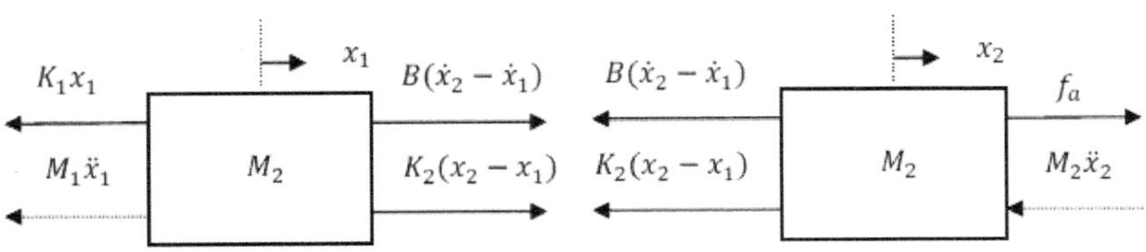

Fig. 2.16 Free-body diagrams for example 2.3

$$\sum_{M_1} f_{ext\,M_1} = M_1 \frac{d^2x_1}{dt^2} + B\left(\frac{dx_1}{dt} - \frac{dx_2}{dt}\right) + K_2(x_1 - x_2) + K_1x_1 = 0$$

$$\sum_{M_2} f_{ext\,M_2} = M_2 \frac{d^2x_2}{dt^2} - B(\frac{dx_1}{dt} - \frac{dx_2}{dt}) - K_2(x_1 - x_2) - f_a(t) = 0$$

In fig. 2.17 and fig. 2.18 the corresponding modeling diagram developed by using both the SIMULINK and SIMSCAPE environments are shown.

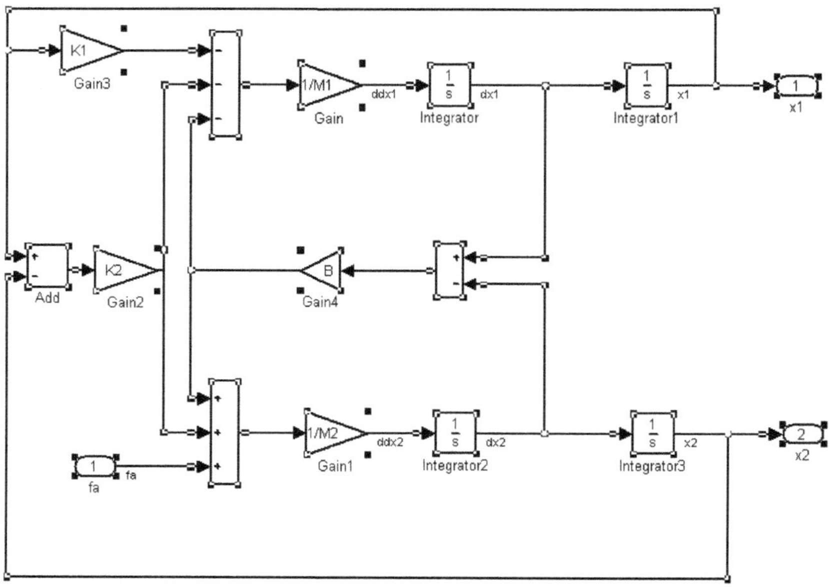

Fig. 2.17 SIMULINK diagram corresponding to example 2.3

The modeling of rotational mechanical systems follow a similar approach to the one discussed for translational mechanical systems.

In table 2.6 the main elements that appear at rotational mechanical systems with its corresponding icons and dynamic equations are described.

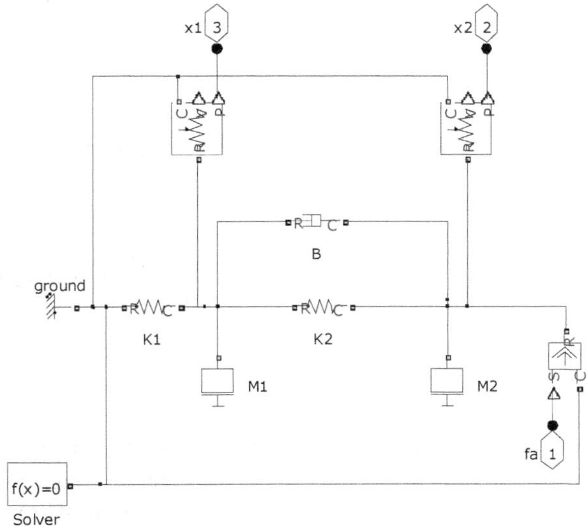

Fig. 2.18 SIMSCAPE diagram corresponding to example 2.3

Table 2.6 Rotational mechanical elements and corresponding dynamic equations

Inertia	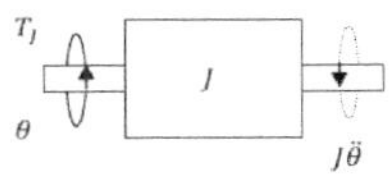	$T_J = J\dfrac{d^2\theta}{dt^2}$
Rotational Spring	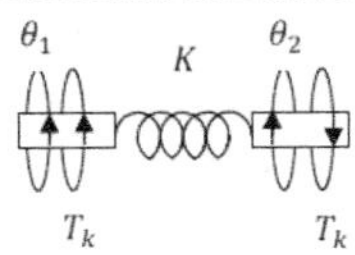	$T_K = K(\theta_1 - \theta_2)$
Rotational Damper	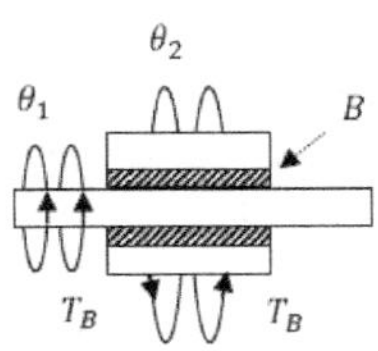	$T_B = B(\dfrac{d\theta_1}{dt} - \dfrac{d\theta_2}{dt})$
Gear	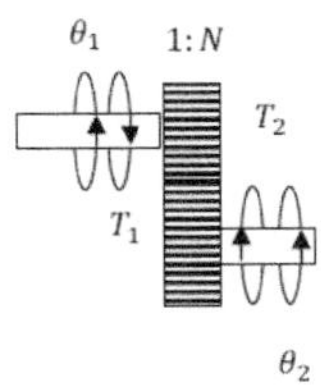	$T_2 = \dfrac{n_2}{n_1}T_1$ $\theta_2 = -\dfrac{n_1}{n_2}\theta_1$

In the same way, the D'Alambert law is applied to get the dynamic equations corresponding to the rotational mechanical systems, just be restating the second law of Newton for rotational movement

$$\sum_i T_{ext_i} = J\frac{d^2\theta_i}{dt^2} \tag{2.3}$$

where the summation includes all the external torques applied to the inertia J. Then, by considering the term $T_J = J\frac{d^2\theta}{dt^2}$ as the inertial torque, it results in the rotational D'Alambert law given by

$$\sum_i T_{ext_i} = 0 \tag{2.4}$$

The dynamic equations are obtained by applying the D'Alembert law to each inertia, using the free-body diagram showing all external torques and the inertial torque with corresponding senses, once the positive sense has been previously defined.

Example 2.4

Derive the dynamic equations corresponding to the rotational mechanical system shown in fig. 2.19, with Ta(t) and TL(t) considered as input and θ1(t) and θ2(t) considered as outputs of the system respectively, also drawing the corresponding SIMULINK and SIMSCAPE modeling diagrams.

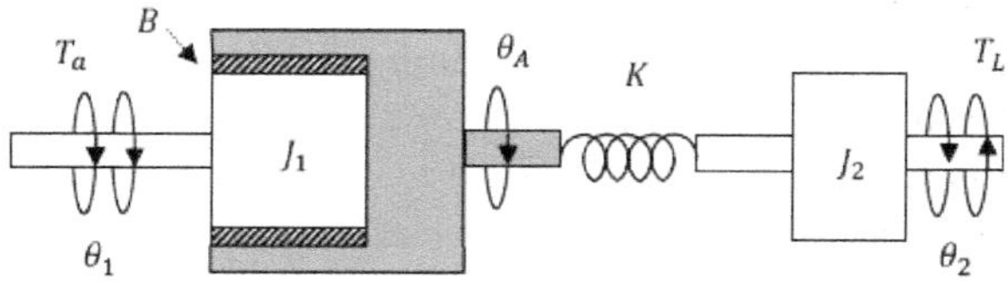

Fig. 2.19 Rotational mechanical system for example 2.4

Solution

Firstly, we draw the free-body diagram for each inertia (fig. 2.20)

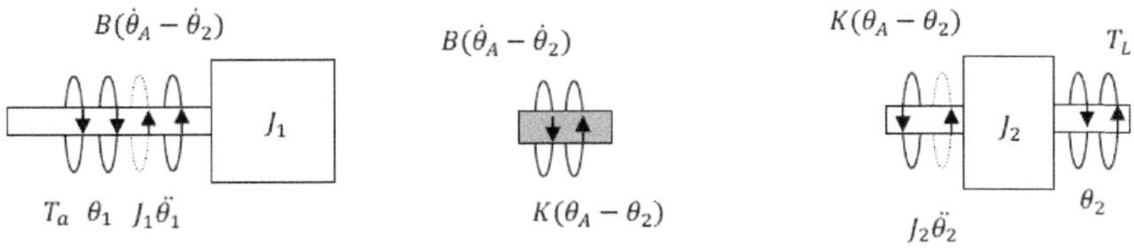

Fig. 2.20 Free-body diagrams for example 2.4

$$\sum_{J_1} T_{ext_{J_1}} = J_1\frac{d^2\theta_1}{dt^2} + B\left(\frac{d\theta_1}{dt} - \frac{d\theta_A}{dt}\right) - T_a(t) = 0$$

$$B\left(\frac{d\theta_1}{dt} - \frac{d\theta_A}{dt}\right) - K(\theta_A - \theta_2) = 0$$

$$\sum_{J_2} T_{ext_{J_2}} = J_2\frac{d^2\theta_2}{dt^2} - K(\theta_A - \theta_2) + T_L(t) = 0$$

Solving $\theta_A(t)$ from the second equation and substituting in the rest, we obtain the dynamic equations.

In fig. 2.21 and fig. 2.22 the corresponding modeling diagram developed by using both the SIMULINK and SIMSCAPE environments are shown

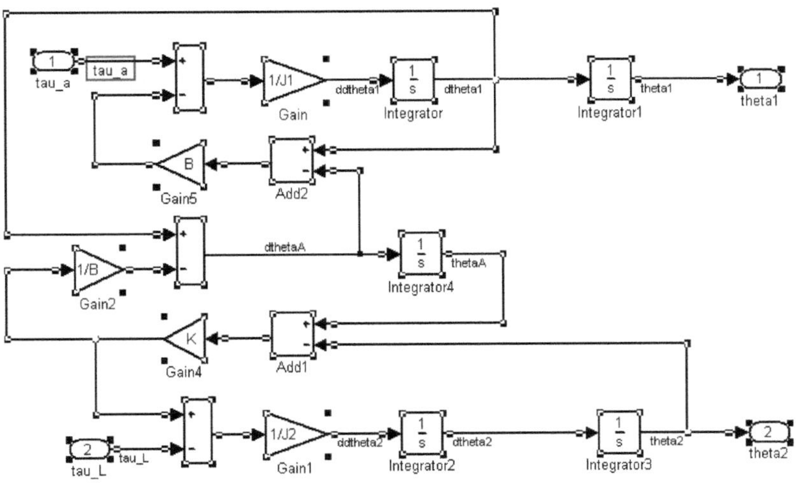

Fig. 2.21 SIMULINK diagram corresponding to example 2.4

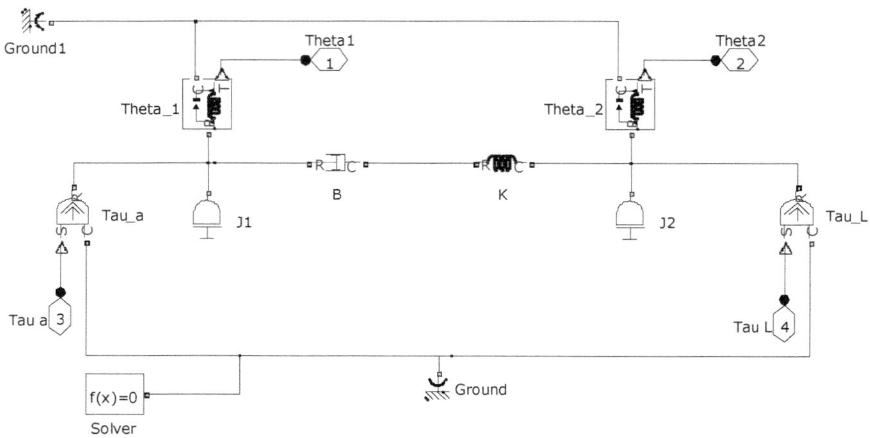

Fig. 2.22 SIMSCAPE diagram corresponding to example 2.4

2.5 Hydraulic Systems

The distributed nature and the nonlinear character of the resistance to flow makes the obtaining of the dynamic equations for the hydraulic systems, difficult in general. In fact, there may be two types of fluid regimes, laminar and turbulent, characterized by linear and nonlinear dynamic equations, respectively.

Table 2.7 Hydraulic elements and corresponding dynamic equations

Pressure Source	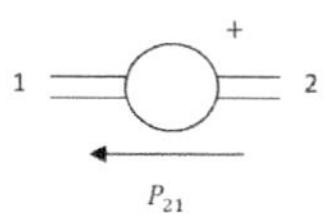	$P_{12} = P_{12}(t)$
Flow Source	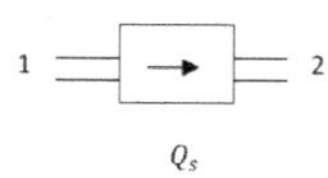	$Q_s = Q_s(t)$
Hydraulic Resistor	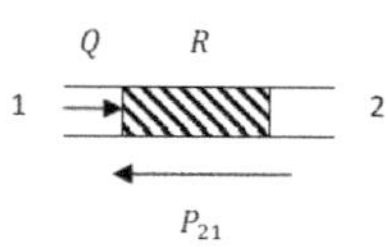	$P_{12} = RQ$ Laminar $P_{12} = C_R Q^n$ Turbulent

Table 2.7 (*continued*)

Hydraulic Inertor	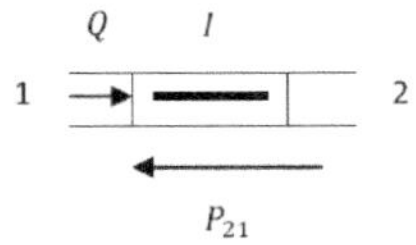	$P_{12} = I\dfrac{dQ}{dt}$
Hydraulic Capacitor (open vessel)	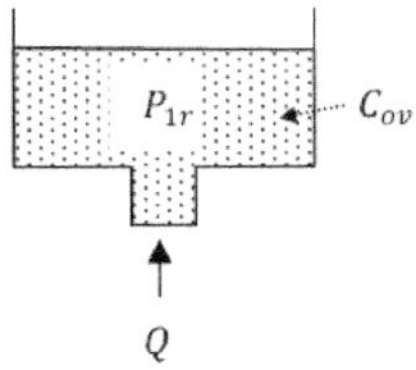	$\dfrac{dP_{1r}}{dt} = \dfrac{1}{C_{ov}}Q$
Hydraulic Capacitor (closed vessel)	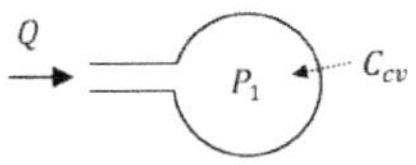	$\dfrac{dP_{1r}}{dt} = \dfrac{1}{C_{cv}}Q$

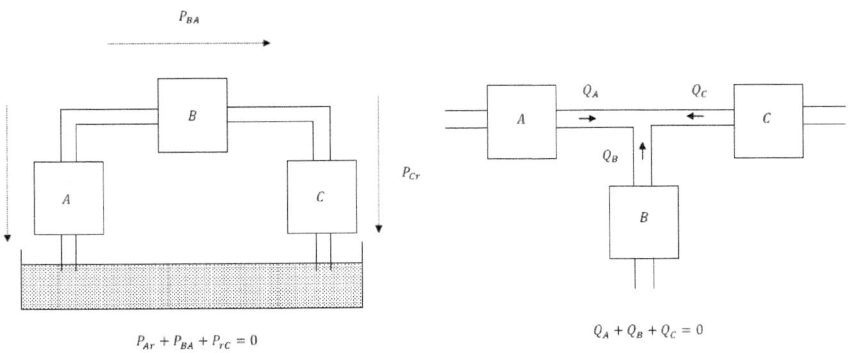

Fig. 2.23 Conservation laws used to obtain the dynamic equations for hydraulic systems.

In order to derive the dynamic equations corresponding to fluid systems, we are going to assume linear and lumped systems in general, made up of combinations of ideal hydraulic elements.

In table 2.7 the main elements that appear at hydraulic systems translational with its corresponding icons and dynamic equations are described.

The fluid conservation laws employed to obtain the complete set of equations involving the hydraulic elements are the same corresponding to Kirchoff's laws already described. The compatibility law settles that the sum of pressure drops around a loop must be equal to zero, while the continuity law states that the sum of the flow rates at an hydraulic junction must be also zero (fig. 2.23).

Example 2.5

Derive the dynamic equations corresponding to the hydraulic circuit shown in fig. 2.24, with Ps(t) and P3r(t) considered as input and output of the system respectively, also drawing the corresponding SIMULINK and SIMSCAPE modeling diagrams.

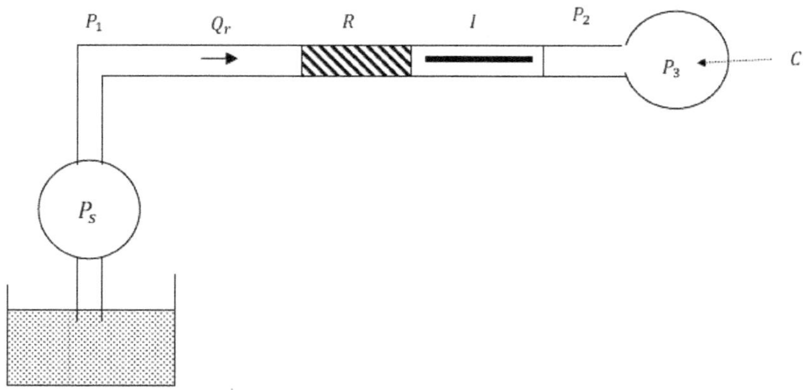

Fig. 2.24 Hydraulic system for example 2.5

Solution

$$P_{3r} + I\frac{dQ_R}{dt} + RQ_R - P_{1r} = 0$$

$$Q_R = C\frac{dP_{3r}}{dt}$$

$$P_{1r} = P_s$$

Solving for Q_R in the second equation and substituting in the first one, yields

$$CI\frac{d^2P_{3r}}{dt^2} + RC\frac{dP_{3r}}{dt} + P_{3r} = P_s$$

In fig. 2.25 and fig. 2.26 the corresponding modeling diagram developed by using both the SIMULINK and SIMSCAPE environments are shown

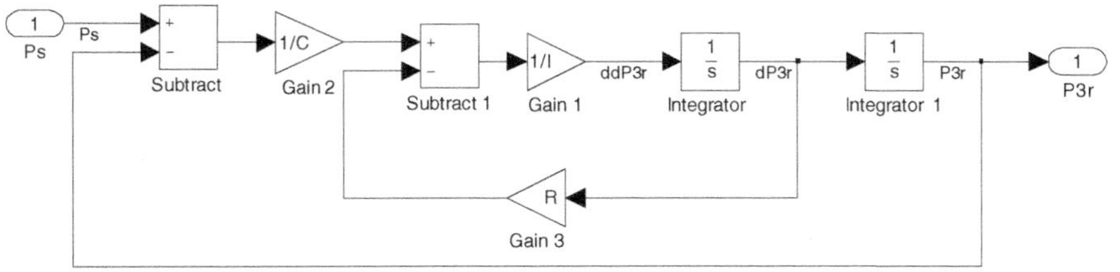

Fig. 2.25 SIMULINK diagram corresponding to example 2.5

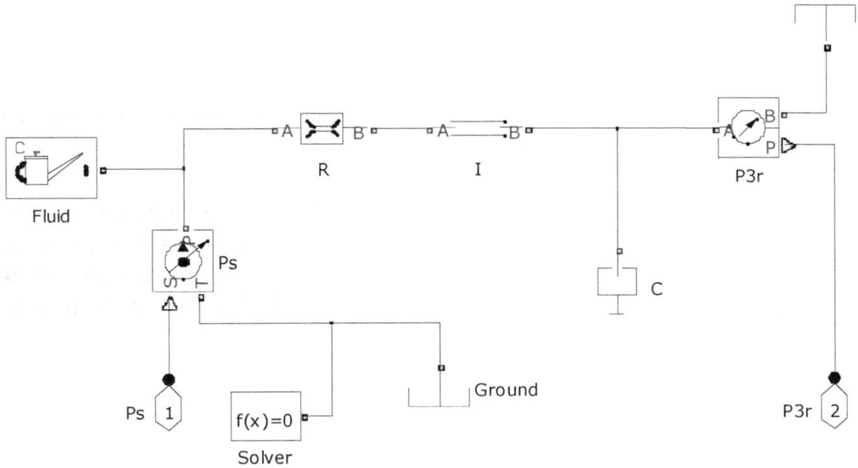

Fig. 2.26 SIMSCAPE diagram corresponding to example 2.5

2.6 Thermal Systems

The mathematical modeling of thermal systems is based on the fundamental laws of the thermodynamics. Thermal systems are in general distributed, so that their dynamic equations are expresses as partial differential equations. We will restrict our analysis to lumped system and approximated linear behavior in general.

Table 2.8 Thermal elements and corresponding dynamic equations

Temperature Source	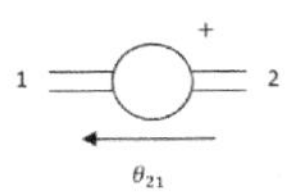	$T_{1r} = T_{1r}(t)$
Heatflow Source	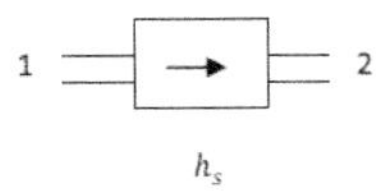	$h = h_s(t)$

Table 2.8 (*continued*)

Conductive Thermal Resistor	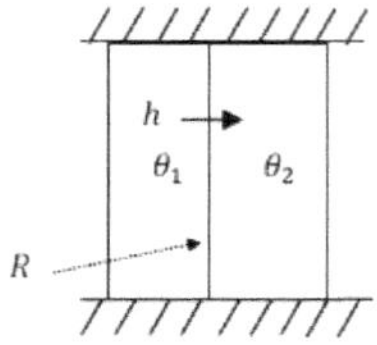	$T_{12} = R_{cond}\, h$ $R_{cond} = \frac{L}{KA}$
Convective Thermal Resistor	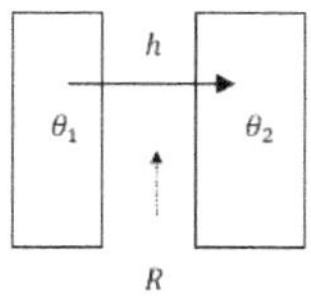	$T_{12} = R_{conv}\, h$ $R_{conv} = \frac{1}{h_c A}$
Radiation Thermal Resistor	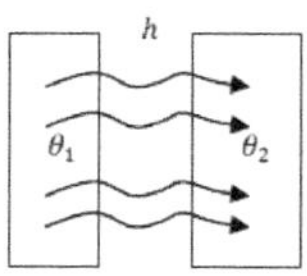	$(T_1^{\,4} - T_2^{\,4}) = \frac{1}{\sigma F_{EF_A A}}\, h$
Thermal Capacitor (Mass)	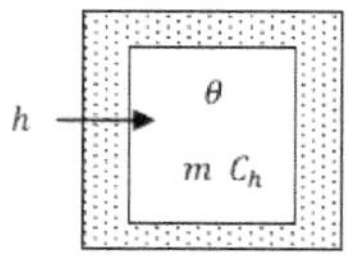	$\frac{dT_{1r}}{dt} = \frac{1}{C_h}\, h$ $C_h = m\, c_p$

In table 2.8 the main elements that appear at thermal systems with its corresponding icons and dynamic equations are described.

The thermal conservation laws employed to obtain the complete set of equations involving the thermal elements are the energy balance equations for each thermal mass, considering the net heat input flow, net heat output flow and heat flow generated, so that

$$\frac{d\theta_{1r}}{dt} C_h = \sum_{m} (q_{in} - q_{out}) + q_{gen} \tag{2.5}$$

Example 2.6

Derive the dynamic equations corresponding to the thermal circuit shown in Fig. 2.27, with q(t) considered as input and θ1(t) and θ2(t) considered as output of

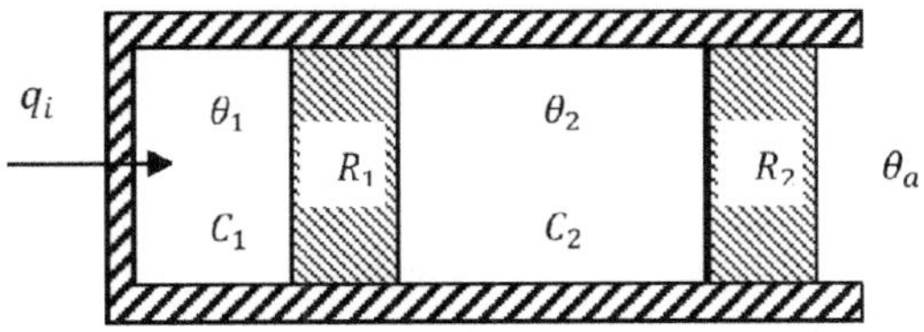

Fig. 2.27 Thermal system for example 2.6

the system respectively, also drawing the corresponding SIMULINK and SIMSCAPE modeling diagrams.

Solution

$$C_1 \frac{d\theta_1}{dt} = q_i(t) - \frac{(\theta_1 - \theta_2)}{R_1}$$

$$C_2 \frac{d\theta_2}{dt} = \frac{(\theta_1 - \theta_2)}{R_1} - \frac{(\theta_2 - \theta_a)}{R_2}$$

In fig. 2.28 and fig. 2.29 the corresponding modeling diagram developed by using both the SIMULINK and SIMSCAPE environments are shown

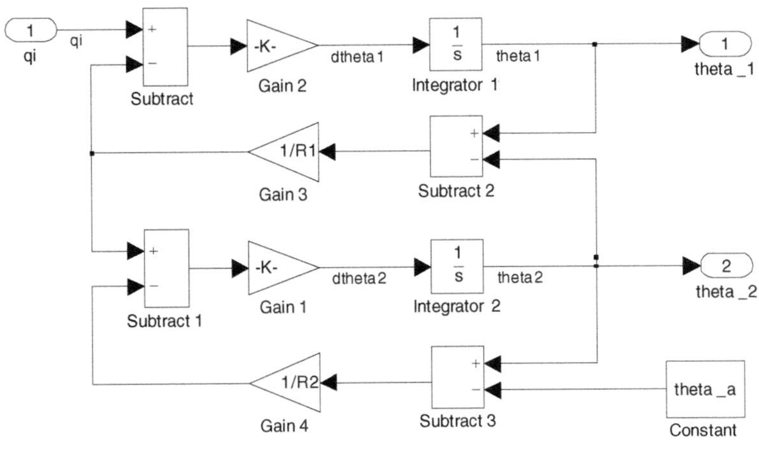

Fig. 2.28 SIMULINK diagram corresponding to example 2.6

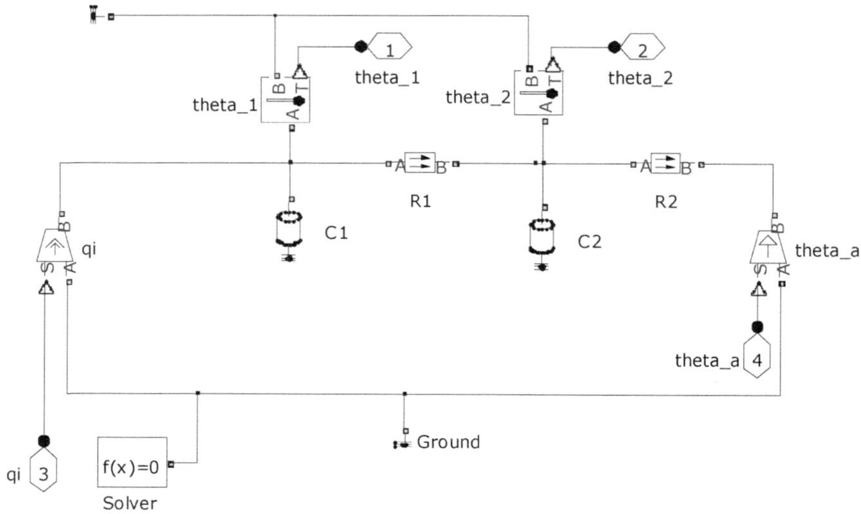

Fig. 2.29 SIMSCAPE diagram corresponding to example 2.6

2.7 Hybrid Systems

Many dynamical systems consist of combinations of electric, mechanical, fluid, and thermal systems, resulting in electromechanical, hydromechanical, and so on. In order to derive the dynamic equations of the combined systems we have to employ coupling devices to convert one kind of magnitude (electrical, mechanical, etc.) to another. In this way, we are going to assume linear and lumped systems in general, made up of combinations of ideal converters.

We will restrict our analysis to electromechanical and hydromechanical converters, both of displacement and rotational types, and signal converting transducers.

In table 2.9 the main converters that appear at electromechanical and hydromechanical systems with its corresponding icons and dynamic equations are described.

In each case, the coupling device converts one kind of energy to another throughout a coupling coefficient α_t and D_t for electromechanical and hydromechanical converters respectively.

The electromechanical translational converter represents an electromagnet, whose primary use is as an electric drive (fig. 2.30).

Table 2.9 Electromechanical and hydromechanical elements with corresponding dynamic equations

Electromechanical Translational Converter	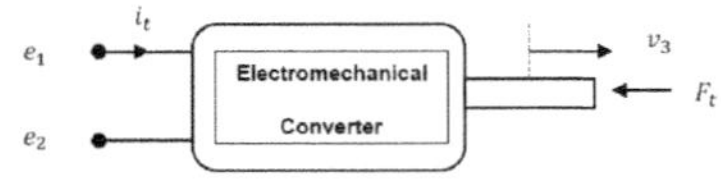	$e_{12} = \alpha_t v_3$ $i_t = \frac{1}{\alpha_t} F_t$
Electromechanical Rotational Converter	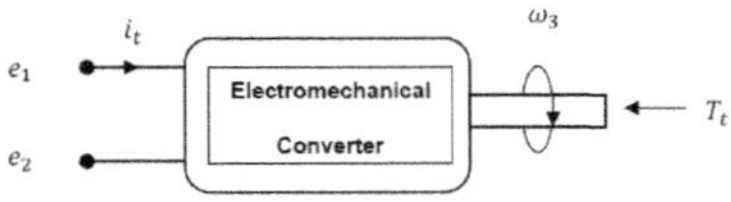	$e_{12} = \alpha_t \omega_3$ $i_t = \frac{1}{\alpha_t} T_t$
Hydromechanical Translational Converter	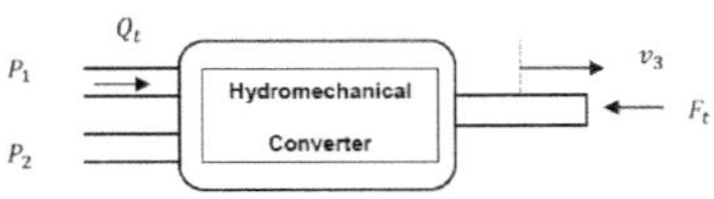	$P_{12} = D_t v_3$ $Q_t = \frac{1}{D_t} F_t$
Hydromechanical Translational Converter	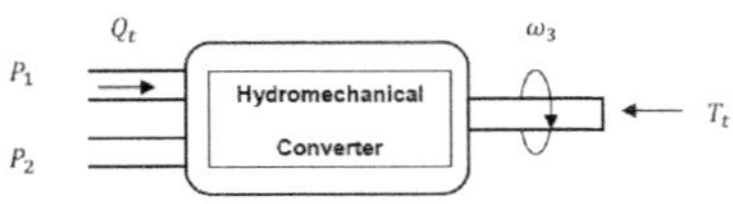	$P_{12} = D_t \omega_3$ $Q_t = \frac{1}{D_t} T_t$

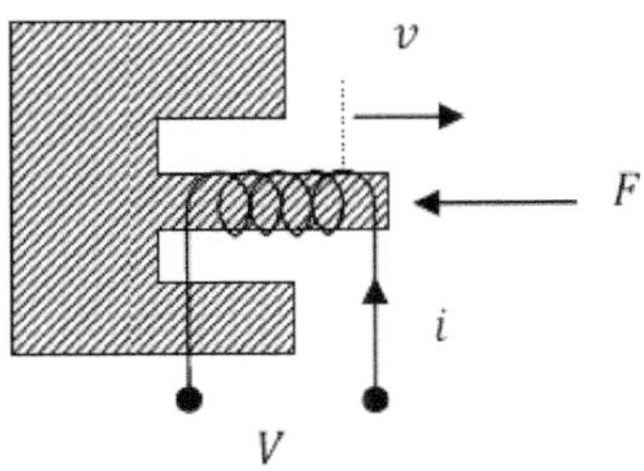

Fig. 2.30 Schematic of the electromagnet device.

The electromechanical rotational converter represents both an ideal DC motor and an ideal DC generator, whose primary use is in DC electric circuits (fig. 2.31).

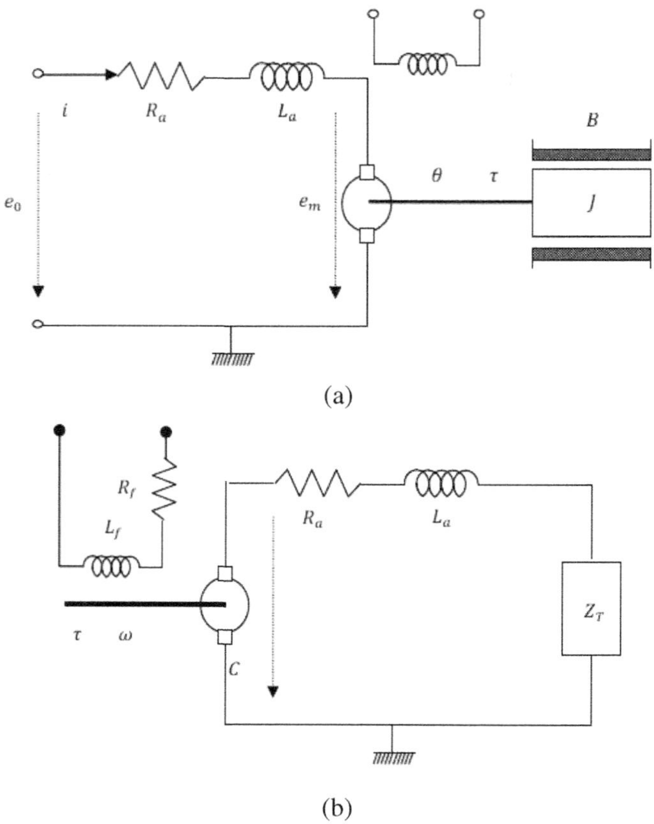

Fig. 2.31 Schematic of the DC motor (a) and DC generator (b) devices.

On the other hand, the hydromechanical translational converter represents an ideal hydraulic piston, whose primary use is as a mechanical drive (fig. 2.32).

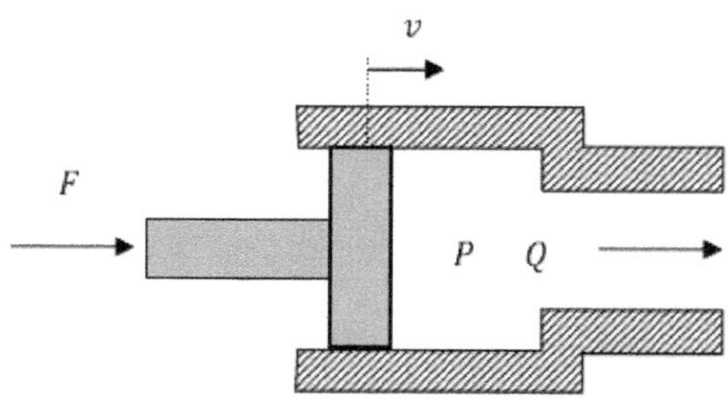

Fig. 2.32 Schematic of the hydraulic piston device.

Meanwhile, the hydromechanical rotational converter represents both an ideal centrifugal pump and an ideal hydraulic turbine, whose primary use is in hydraulic drive systems (fig. 2.33).

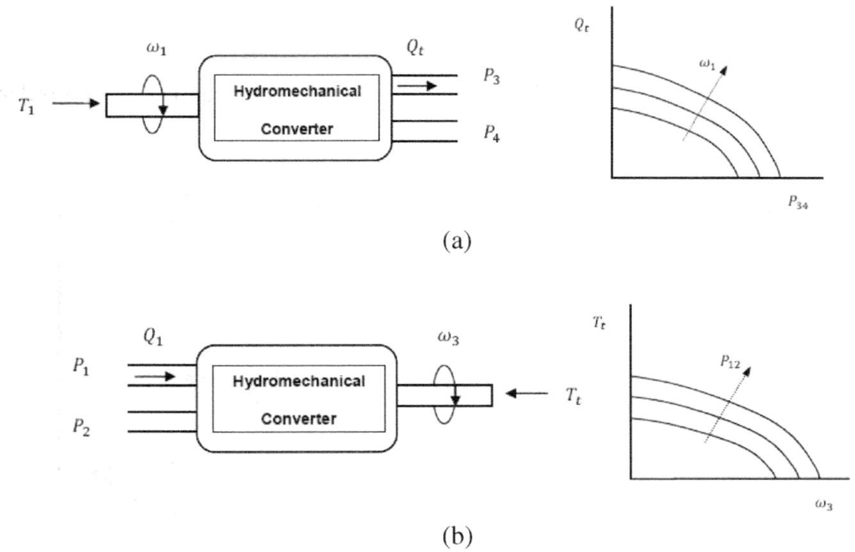

Fig. 2.33 Schematic of the centrifugal pump (a) and hydraulic turbine (b) devices.

Table 2.10 Some signal converting transducers with corresponding dynamic equations

Translational Potentiometer	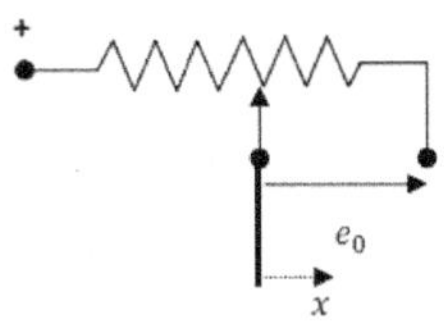	$e_0 = \frac{E}{x_{max}} x(t)$
Rotational Potentiometer	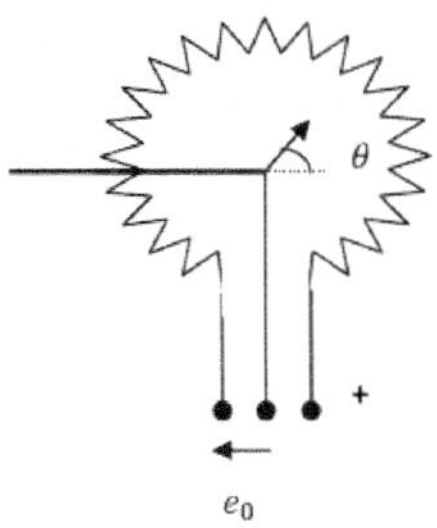	$e_0 = \frac{E}{\theta_{max}} \theta(t)$

Table 2.10 (*continued*)

Tachometer	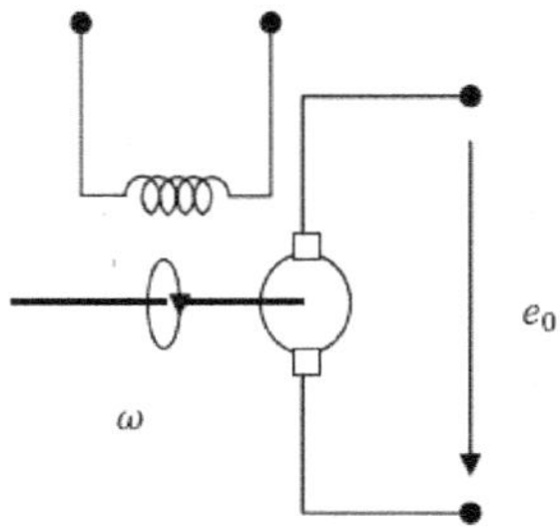	$v_0 = k\,\omega$

Besides this, also to be included are the signal converting transducers, characterized for being energy converters that have a negligible load, such as potentiometers, and tachometers among others. In table 2.10 these converting transducers with its corresponding icons and dynamic equations are described.

Example 2.7

Derive the dynamic equations corresponding to the electromechanic circuit shown in fig. 2.34, with $e_i(t)$ considered as input and x(t) and i(t) considered as output of the system respectively, also drawing the corresponding SIMULINK and SIMSCAPE modeling diagrams.

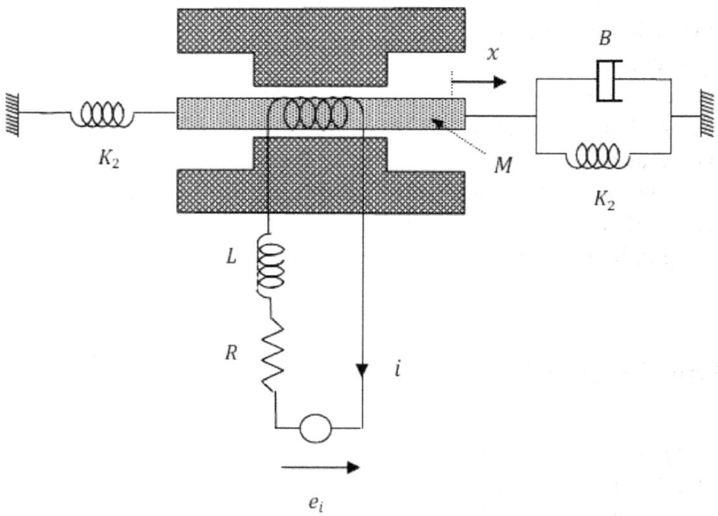

Fig. 2.34 Hybrid electromechanic system for example 2.7.

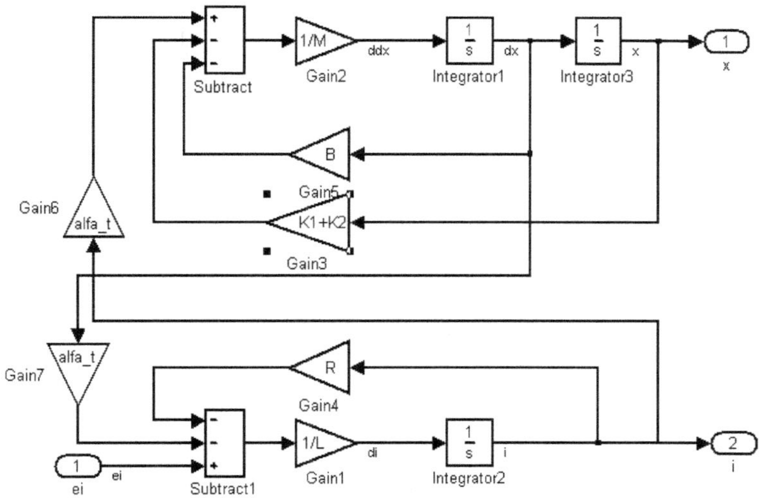

Fig. 2.35 SIMULINK diagram corresponding to example 2.7

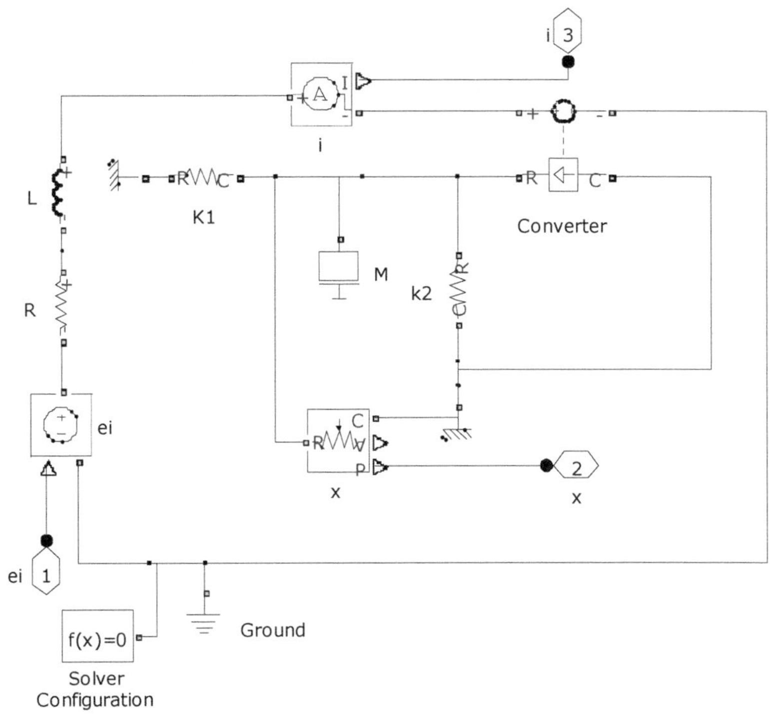

Fig. 2.36 SIMSCAPE diagram corresponding to example 2.7

Solution

$$M\frac{d^2x}{dt^2} + B\frac{dx}{dt} + (K_1 + K_2)x = \alpha_t\, i$$

$$L\frac{di}{dt} + R\, i + \alpha_t\frac{dx}{dt} = e_i(t)$$

In fig. 2.35 and fig. 2.36 the corresponding modeling diagram developed by using both the SIMULINK and SIMSCAPE environments are shown.

Chapter 3
System Description

This chapter deals with the description of continuous-time systems, in particular, the external and internal description of linear and time invariant (LTI) type, and introduces the linearization technique employed for approximating nonlinear system behavior.

The Laplace transform method to find the analytical response of LTI systems to different input signals is also described. We also derive the transfer function as a mean to represent LTI systems in the Laplace domain, both for univariable and multivariable systems. The state space approach and its relation to the transfer function are also treated.

Finally, we describe the techniques to obtain block diagrams starting from dynamical equations for LTI systems, and the block reduction techniques for obtaining a unique transfer function representing the system behavior. Accompanying the subject matter, some applications are illustrated in MATLAB code to improve the concepts described.

3.1 Continuous Time Systems

A continuous-time system is one for which the inputs, state variables and outputs are defined over some continuous range of time, and its behavior is described by differential equations.

Some of the input signals are used as test signals both to analyze the system behavior or to design control systems. These inputs are the impulse, step, ramp, or sinusoidal as is shown in table 3.1.

In general, a continuous system with r inputs and s outputs (fig. 3.1) can be described by s independent input-output differential equations

$$\begin{aligned}
&f_1\left(y_1^{n)},\dots,\dot{y}_1,y_1,\dots,y_s^{n)},\dots,\dot{y}_s,y_s,u_1^{m)},\dots,\dot{u}_1,u_1,\dots,u_r^{m)},\dots,\dot{u}_r,u_r,t\right)=0\\
&f_2\left(y_1^{n)},\dots,\dot{y}_1,y_1,\dots,y_s^{n)},\dots,\dot{y}_s,y_s,u_1^{m)},\dots,\dot{u}_1,u_1,\dots,u_r^{m)},\dots,\dot{u}_r,u_r,t\right)=0\\
&\vdots\\
&f_s\left(y_1^{n)},\dots,\dot{y}_1,y_1,\dots,y_s^{n)},\dots,\dot{y}_s,y_s,u_1^{m)},\dots,\dot{u}_1,u_1,\dots,u_r^{m)},\dots,\dot{u}_r,u_r,t\right)=0
\end{aligned} \tag{3.1}$$

where $m \le n$ for causal behavior, with $f_i, i = 1, \dots, s$ non-linear in general. The parameter n is denoted as the system order, and plays a fundamental role for the system behavior.

Table 3.1 Some test input signals used for system analysis

Impulse	$\delta(t)$	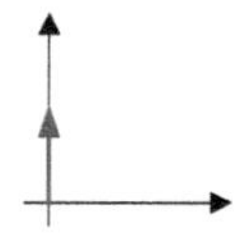
Step	$u_s(t)$	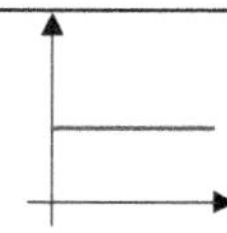
Ramp	$r(t) = t\, u_s(t)$	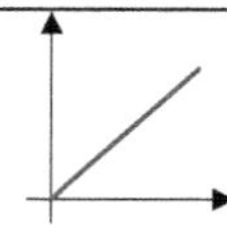
Senoidal	$\sin(\omega t)\, u_s(t)$	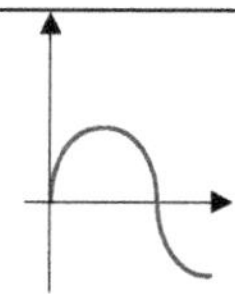

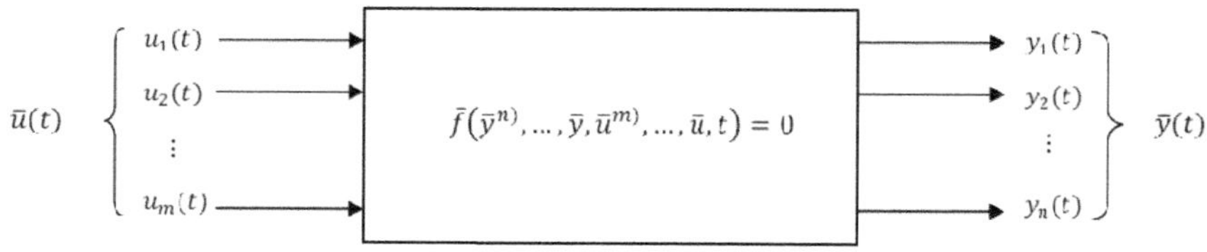

Fig. 3.1 Continuous multi-input, multi-output system.

If the system is single-input and single-output, the continuous dynamic differential equation is given by

$$f\left(y^{n)}, \ldots, \dot{y}, y, u^{m)} \ldots, \dot{u}, u, t\right) = 0 \tag{3.2}$$

where $m \leq n$ for causal behavior, with f non-linear in general.

Analytical methods for exact solving (3.1) given the input signals and an appropriate set of initial conditions are practical only for low order and system dynamics described rather by (3.2). Nevertheless, the application of numerical methods will enable to deal with high order and multi-input, multi-output to obtain approximate solutions.

Let us particularize the analysis for a single input single output system, with f described by a linear combination of both input and output signals together with its derivatives, assuming also the system is stationary, and no direct dependence with time t is made explicit.

3.2 Linear and Invariant Time Systems

It is said that a system is linear when it fulfills the principle of superposition, which states that the system's response to a set of simultaneous inputs can be decomposed into the sum of individual responses (fig. 3.2). On the contrary, nonlinear systems are those which superposition does not hold.

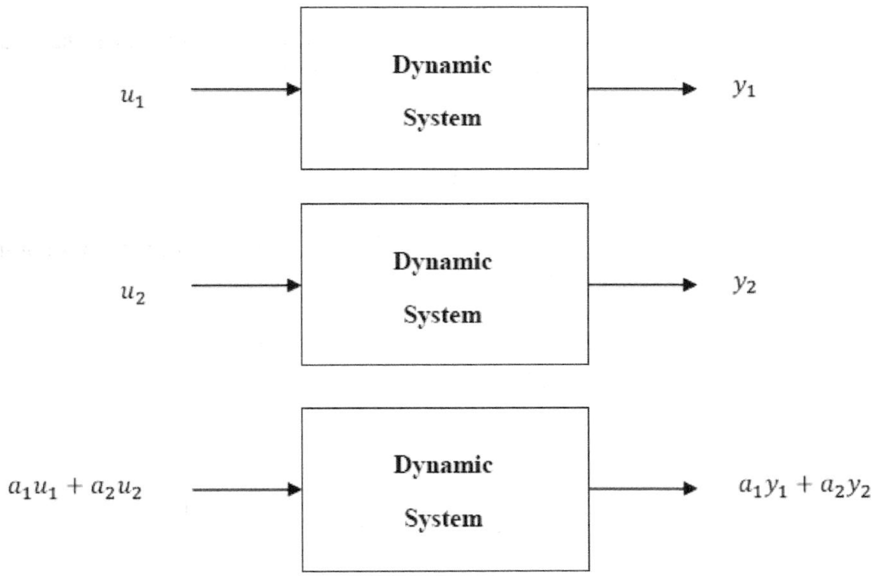

Fig. 3.2 The application of the principle of superposition

A continuous-time system defined by differential equations is said to be linear if it can be expressed as a linear combination of derivatives of the output and input in the form

$$a_n y^{n)} + \cdots + a_1\dot{y} + a_0 y = b_m\, u^{m)} + \cdots + b_1\dot{u} + b_0 u \tag{3.3}$$

where the parameters a_j, $j = 1 \dots n$ and b_k, $k = 1 \dots m$ are constants or functions of time t. In the event that these parameters do not depend explicitly on the time t, the system is said to be also time-invariant, and if it is not the case, the system is said to be nonlinear.

Nearly all systems are inherently nonlinear in nature, since most of the relationships in physics are nonlinear. Moreover, most of the linear systems are a special case of nonlinear systems in limited ranges of operation. Besides this, the nonlinear nature of the physical elements constituting the system may be an essential feature of the system, causing the overall system behavior to be nonlinear as with the dead zone, relay, saturation, etc. characteristics (fig. 3.3).

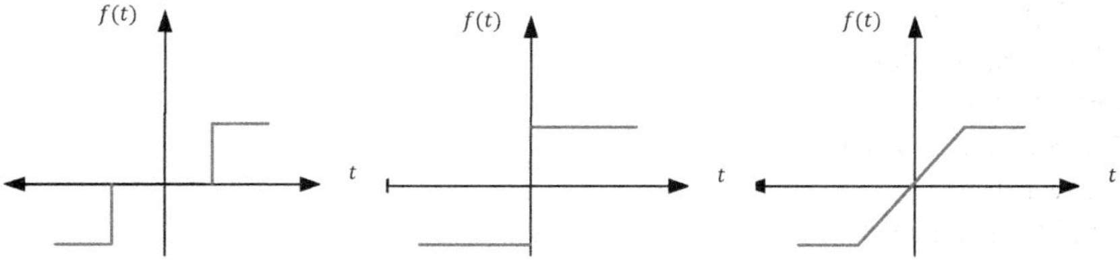

Fig. 3.3 Some examples of nonlinear relationships

3.3 Systems Linearization

The objective of the linearization technique is to derive a linear approximate system whose response will agree closely with that of the original nonlinear system which it comes from.

In order to apply linearization method, we use the disturbance method, which consists in considering small variations around a nominal operating point, making into the non-linear relationship $y = f(u)$, the following substitutions:

1. The independent variable u is replaced by $u = u_0 + \delta u$, u_0 being the operating point of u, and δu the disturbance applied on u.
2. The dependent variable or curve representing the non-linearity $y = f(u)$ is replaced by its tangent at the nominal operating point u_0 (fig. 3.4).

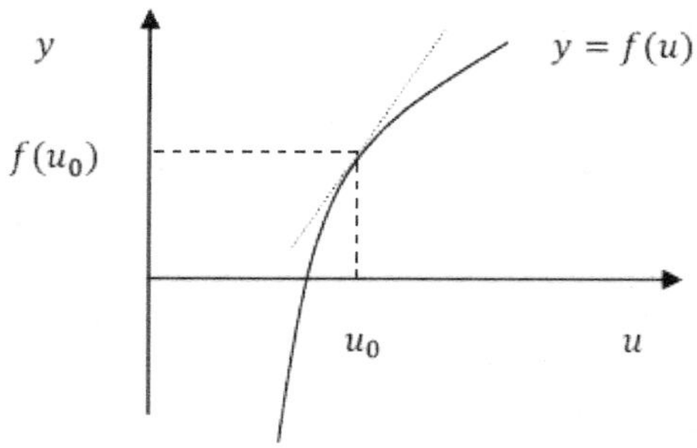

Fig. 3.4 Linearization of $y = f(u)$ at operation point u_0

Approximating $y = f(u)$ by the Taylor's expansion we have

$$y = f(u) = f(u_0 + \delta u) = f(u_0) + \dot{f}(u_0)\delta u + \frac{1}{2!}\ddot{f}(u_0)\delta^2 u + \cdots \tag{3.4}$$

Approaching the nonlinear relation by its first derivative (tangent of the curve) it follows that

$$y = f(u) = f(u_0 + \delta u) \cong f(u_0) + \dot{f}(u_0)\delta u \tag{3.5}$$

which represents the substitution to be made on the dependent variable y whenever $u = u_0 + \delta u$.

By extending this method to a multivariable nonlinear relationship $y = f(u_1, u_2, \ldots, u_n)$ in general, we have

$$\begin{aligned} y = f(u) = f(u_{10} + \delta u_1, \ldots, u_{n0} + \delta u_n) \cong \\ f(u_{10}, \ldots, u_{n0}) + \frac{\partial f}{\partial u_1}\delta u_1 + \cdots + \frac{\partial f}{\partial u_n}\delta u_n \end{aligned} \tag{3.6}$$

where $\bar{u}_0 = (u_{10}, u_{20}, \ldots, u_{n0})$ the nominal operating point, where the partial derivatives of f w.r.t. to u_i are evaluated at $\bar{u}_0$.

In case nonlinear system relationship is given by a set of differential equations, the disturbance method in implicit form will be applied to

$$F\left(y^{n)}, \ldots, \dot{y}, y, u^{m)}, \ldots, \dot{u}, u\right) = 0 \tag{3.7}$$

Therefore, following the same approach used in (3.6), we have

$$\begin{aligned} F\left(y^{n)}, \ldots, \dot{y}, y, u^{m)}, \ldots, \dot{u}, u\right) = F\left({y_0}^{n)}, \ldots, \dot{y}_0, y_0, {u_0}^{m)}, \ldots, \dot{u}_0, u_0\right) + \\ + \frac{\partial F}{\partial y^{n)}}\delta y^{n)} + \cdots + \frac{\partial F}{\partial \dot{y}}\delta \dot{y} + \frac{\partial F}{\partial y}\delta y + \frac{\partial F}{\partial u^{m)}}\delta u^{m)} + \cdots + \frac{\partial F}{\partial \dot{u}}\delta \dot{u} \\ + \frac{\partial F}{\partial y}\delta u \end{aligned} \tag{3.8}$$

The nominal operating point $\bar{u}_0 = (y_0, u_0)$ around which we will make the linearization, is necessarily the solution of the differential equation describing the system dynamics, and is derived from equation (3.7), making zero, the derivatives of the implicit differential equation at $u = u_0$ in

$$F\left(y_0{}^{n)}, \ldots, \dot{y}_0, y_0, u_0{}^{m)}, \ldots, \dot{u}_0, u_0\right) = 0 \tag{3.9}$$

In this way, the second term on the right hand side of (3.8) is the differential equation linearized around the nominal operating point $\bar{u}_0 = (y_0, u_0)$, that is approximated by

$$\frac{\partial F}{\partial y^{n)}}\delta y^{n)} + \cdots + \frac{\partial F}{\partial y}\delta y + \frac{\partial F}{\partial u^{m)}}\delta u^{m)} + \cdots + \frac{\partial F}{\partial y}\delta u = 0 \tag{3.10}$$

where the partial derivatives of F w.r.t. to each independent variables of (3.7) are evaluated at $\bar{u}_0$, and whose validity is restricted to a small region $\delta\bar{u} = \bar{u} - \bar{u}_0$ around $\bar{u}_0$.

Example 3.1

The diagram of (fig. 3.5) represents a two tank in series system, whose dynamics is given by the set of nonlinear differential equations

$$A_1\frac{dh_1}{dt} = f_1(h_1, h_2, q_e) = q_e - K_1\sqrt{h_1 - h_2}$$

$$A_2\frac{dh_2}{dt} = f_2(h_1, h_2, q_e) = K_1\sqrt{h_1 - h_2} - K_2\sqrt{h_2}$$

where h_1, h_2, and q_e respectively are the height of tank 1, tank 2 and input flow to tank 1, while the rest of parameters A_1, A_2, K_1, and K_2 stand for the tank areas and the discharge coefficient.

Obtain the linearized system around the operation point $q_e = q_{e0}$.

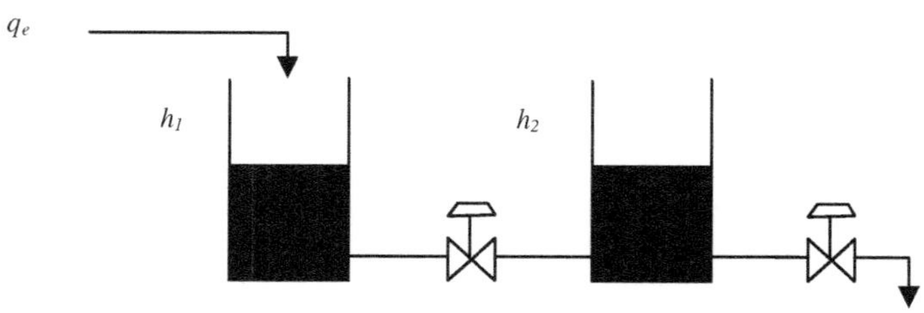

Fig. 3.5 Linearization of a two tank system.

Solution

In first place we must express the system equations in implicit form, so as to derive the linearized approximated system, calculating afterwards the operation point $\bar{u}_0 = (h_{10}, h_{20}, q_{e0})$.

$$F_1(\dot{h}_1, h_1, \dot{h}_2, h_2, q_e) = A_1 \frac{dh_1}{dt} - q_e + K_1\sqrt{h_1 - h_2} = 0$$

$$F_2(\dot{h}_1, h_1, \dot{h}_2, h_2, q_e) = A_2 \frac{dh_2}{dt} - K_1\sqrt{h_1 - h_2} + K_2\sqrt{h_2} = 0$$

Making zero the derivatives in both equations we have a system equation for solving h_{10} and h_{20} as a function of q_{e0}, so we have

$$K_1\sqrt{h_{10} - h_{20}} = q_{e0}$$

$$-K_1\sqrt{h_{10} - h_{20}} + K_2\sqrt{h_{20}} = 0$$

Now, the partial derivatives of F_1 and F_2 w.r.t. to the dependent variables at $\bar{u}_0 = (h_{10}, h_{20}, q_{e0})$ are calculated, that is,

$$\frac{\partial F_1}{\partial \dot{h}_1} = A_1 \qquad \frac{\partial F_2}{\partial h_2} = \frac{K_1}{2\sqrt{h_{10} - h_{20}}} + \frac{K_2}{2\sqrt{h_{20}}}$$

$$\frac{\partial F_1}{\partial h_1} = \frac{K_1}{2\sqrt{h_{10} - h_{20}}} \qquad \frac{\partial F_2}{\partial \dot{h}_1} = 0$$

$$\frac{\partial F_1}{\partial \dot{h}_2} = 0 \qquad \frac{\partial F_2}{\partial h_1} = -\frac{K_1}{2\sqrt{h_{10} - h_{20}}}$$

$$\frac{\partial F_1}{\partial h_2} = -\frac{K_1}{2\sqrt{h_{10} - h_{20}}} \qquad \frac{\partial F_2}{\partial \dot{h}_2} = A_2$$

$$\frac{\partial F_1}{\partial q_e} = -1 \qquad \frac{\partial F_2}{\partial q_e} = 0$$

Thus, defining the incremental variables as $\delta h_1 = h_1 - h_{10}$, $\delta h_2 = h_2 - h_{20}$, and $\delta q_e = q_e - q_{e0}$, we have the linearized equation set

$$A_1 \frac{d\delta h_1}{dt} = \delta q_e - \frac{K_1}{2\sqrt{h_{10} - h_{20}}}(\delta h_1 - \delta h_2)$$

$$A_2 \frac{d\delta h_2}{dt} = \frac{K_1}{2\sqrt{h_{10} - h_{20}}}(\delta h_1 - \delta h_2) - \frac{K_2}{2\sqrt{h_{20}}}\delta h_2$$

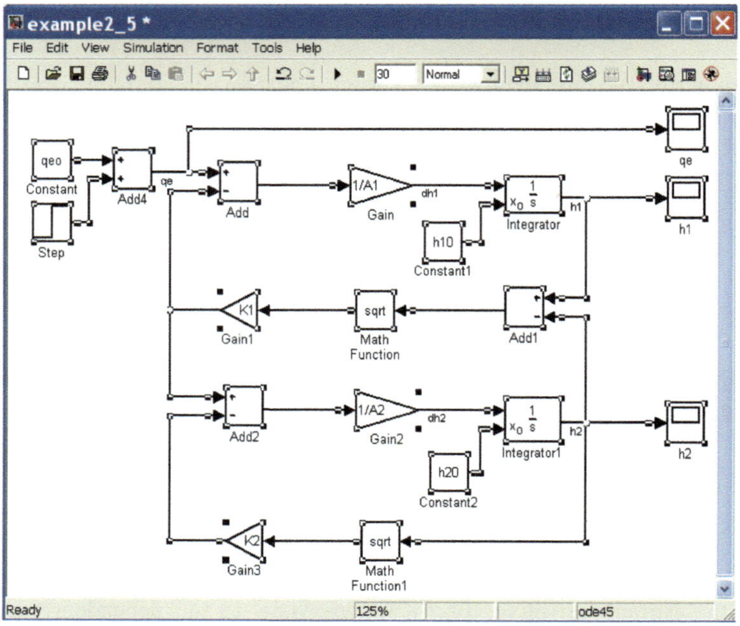

Fig. 3.6 SIMULINK diagram corresponding to the nonlinear two tank system.

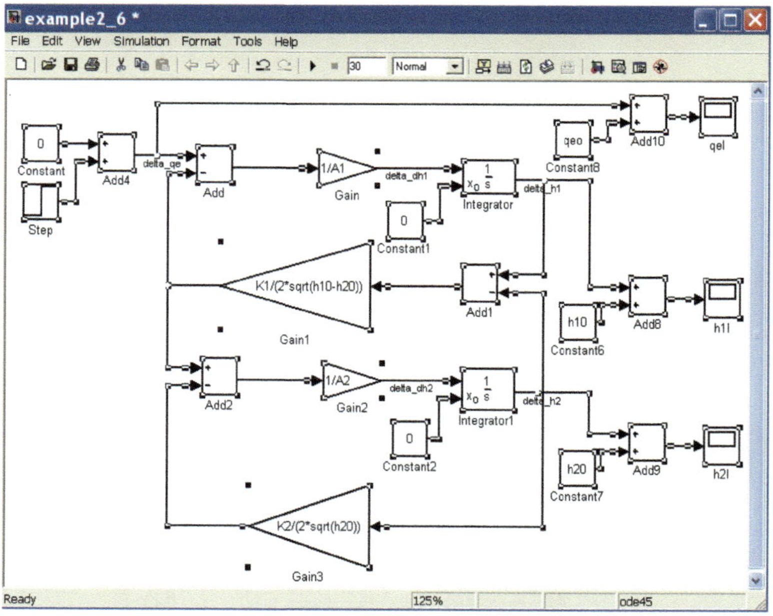

Fig. 3.7 SIMULINK diagram corresponding to the linearized two tank system.

In fig. 3.6 we show the SIMULINK diagrams corresponding to the nonlinear two tank system and its linearized approximation (fig. 3.7), together with their responses to pulse step input with increasing amplitude. For this experience, we have selected an operating point $\bar{u}_0 = (h_{10}, h_{20}, q_{e0}) = (1.25,1.0,1.0)$ after solving the linearization point equations for $q_{e0} = 1.0.$, with $A_1 = 2.0$, $A_2 = 3.0$, $K_1 = 2.0$ and $K_2 = 1.0$, using an incremental amplitude for q_e from 1.0 to 2.0.

We can observe that as long as the incremental input amplitude δq_e remains closer to the operation point, no significant differences are observed between the nonlinear and linearized responses to pulse changes (fig. 3.8).

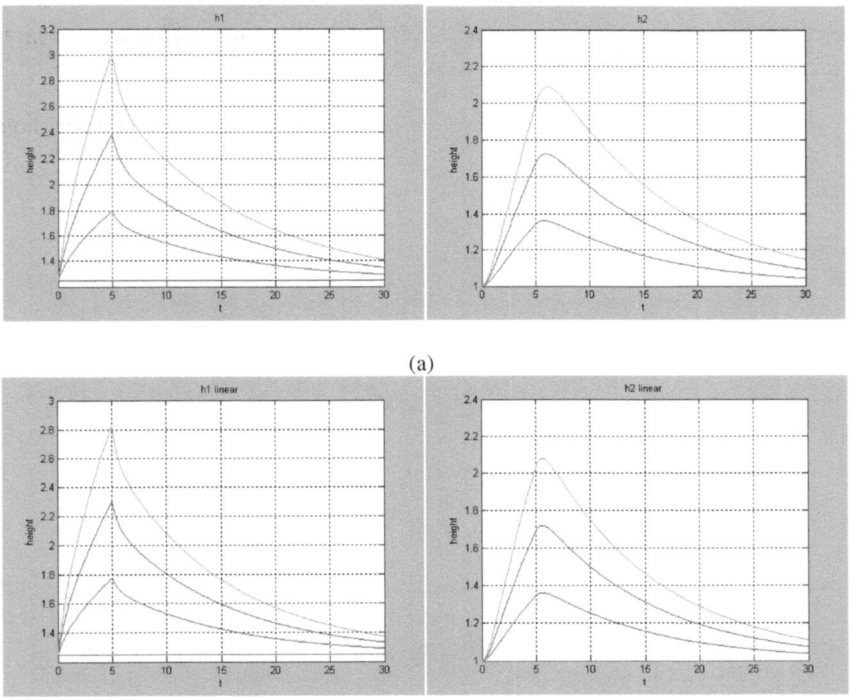

(a)

(b)

Fig. 3.8 Response to incremental pulse changes in q_e corresponding both to the nonlinear (a) and to the linearized (b) two tank system.

3.4 Laplace Transform

The technique of Laplace transform is used to solve the linear differential equations with constant coefficients expressed in the time domain which define the linear system behavior, thus transforming them into linear algebraic equations expressed in a complex domain.

3.4.1 Direct Laplace Transform

The Laplace transform of $f(t)$ is represented symbolically as $\mathcal{L}(f(t))$ and is defined as

$$\mathcal{L}(f(t)) = F(s) = \int_0^\infty f(t)e^{-st}\,dt \tag{3.11}$$

where $s = \sigma + j\omega$ is a complex variable.

Applying equation (3.11) it is possible to derive a pair $f(t) \underset{\mathcal{L}}{\Rightarrow} F(s)$ for each function, so that it is possible to derive a complete table for the most representative input and output signals appearing in dynamic continuous system.

Example 3.2

Derive the Laplace transform of the impulse, step and ramp signals.

Solution

1. $\mathcal{L}(\delta(t)) = \int_0^\infty \delta(t)e^{-st}\,dt = 1$, therefore $\delta(t) \underset{\mathcal{L}}{\Rightarrow} 1$

2. $\mathcal{L}(u_s(t)) = \int_0^\infty u_s(t)e^{-st}\,dt = \int_0^\infty e^{-st}\,dt = \frac{1}{s}$, therefore $u_s(t) \underset{\mathcal{L}}{\Rightarrow} \frac{1}{s}$

3. $\mathcal{L}(r(t)) = \int_0^\infty tu_s(t)e^{-st}\,dt = \int_0^\infty te^{-st}\,dt$, applying the integration by parts, $\mathcal{L}(r(t)) = \frac{1}{s}\int_0^\infty e^{-st}\,dt = \frac{1}{s^2}$, therefore $r(t) \underset{\mathcal{L}}{\Rightarrow} \frac{1}{s^2}$

Nevertheless, in order to find the transform pairs for complex signals and to solve for the responses of dynamic linear systems given by (3.3), we illustrate some useful properties of the Laplace transform:

a) Linearity

$$\mathcal{L}(a_1 f_1(t) + a_2 f_2(t)) = a_1 F_1(s) + a_1 F_2(s) \tag{3.12}$$

b) Damping

$$\mathcal{L}(e^{-at} f(t)) = F(s+a) \tag{3.13}$$

c) Displacement in time

$$\mathcal{L}(f(t-T)u_s(t-T)) = e^{-sT} F(s) \tag{3.14}$$

Table 3.2 Laplace transforms table

Function f(t)	*Laplace Transform F(s)*
$\delta(t)$	1
$u_s(t)$	$\frac{1}{s}$
$t\,u_s(t)$	$\frac{1}{s^2}$
$e^{-at}\,u_s(t)$	$\frac{1}{s+a}$
$te^{-at}\,u_s(t)$	$\frac{1}{(s+a)^2}$
$sen(\omega t)\,u_s(t)$	$\frac{1}{s^2+\omega^2}$
$cos(\omega t)\,u_s(t)$	$\frac{s}{s^2+\omega^2}$
$t^n\,u_s(t)$	$\frac{n!}{s^{n+1}}$
$t^n\,e^{-at}u_s(t)$	$\frac{n!}{(s+a)^{n+1}}$
$e^{-at}sen(\omega t)\,u_s(t)$	$\frac{\omega}{(s+a)^2+\omega^2}$
$e^{-at}cos(\omega t)\,u_s(t)$	$\frac{(s+a)}{(s+a)^2+\omega^2}$

d) Differentiation

$$\mathcal{L}\left(f^{\,n)}(t)\right) = s^n F(s) - s^{n-1} f(0) - s^{n-2}\dot{f}(0) - \cdots - f^{n-1)}(0) \qquad (3.15)$$

being $f(0), \dot{f}(0), \ldots, f^{n-1)}(0)$ the set of initial conditions for $f(t)$.

e) Integration

$$\mathcal{L}\left(\int_0^t f(\tau)d\tau\right) = \frac{1}{s}F(s) \tag{3.16}$$

f) Multiplication by time

$$\mathcal{L}(t^n f(t)) = (-1)^n F^{n)}(s) \tag{3.17}$$

g) Multiplication

$$\mathcal{L}(f_1(t)\, f_2(t)) \neq F_1(s)\, F_2(s) \tag{3.18}$$

h) Final Value

$$f(\infty) = \lim_{t\to\infty} f(t) = \lim_{s\to 0} s\, F(s) \tag{3.19}$$

Then, by combining the Laplace transform of simple signals with the properties described above, it is possible to build a table of Laplace transforms to derive the transform of common functions which appear frequently when the system response is demanded (table 3.2)

Example 3.3

Derive the step response $Y(s)$ corresponding to the mechanical system shown in fig. 3.9 whose dynamic equation is given by

$$M\frac{d^2y}{dt^2} + D\frac{dy}{dt} + Ky = F$$

for values of $M = 1$, $D = 6$ and $K = 9$, starting from initial conditions $y(0) = 0$ and $\frac{dy}{dt}(0) = 0$.

Solution

Applying the Laplace transform properties we get

$$Ms^2\, Y(s) + Ds\, Y(s) + KY(s) = F(s)$$

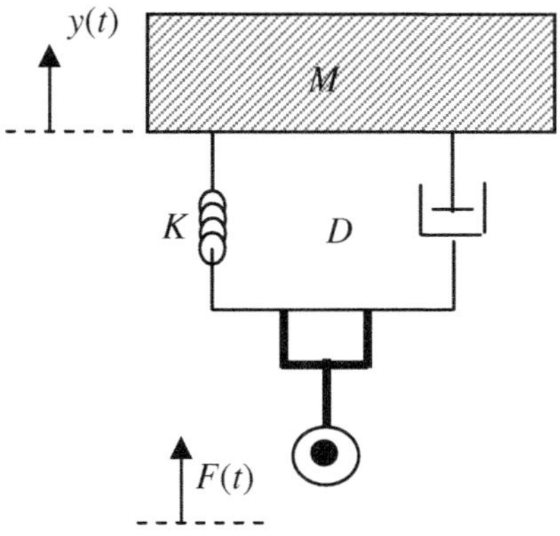

Fig. 3.9 Diagram of the mechanical system for example 3.3.

so that the system response is given by

$$Y(s) = \frac{1}{Ms^2 + Ds + K} F(s)$$

In case of step input $F(s) = \frac{1}{s}$, so that the system response $Y(s)$ particularized for M, D, and K is given by

$$Y(s) = \frac{1}{s(s^2 + 6s + 9)}$$

3.4.2 Inverse Laplace Transform

Once we get the Laplace transform $Y(s)$ of the response of a continuous linear and invariant system as described by (3.3), it is necessary to derive the temporal response $y(t)$ by applying the inverse Laplace transform, which enables to restore the time response format.

The inverse Laplace transform of $Y(s)$ is represented symbolically as $\mathcal{L}^{-1}(Y(s))$ and is defined as the integral in the complex domain

$$\mathcal{L}^{-1}\big(Y(s)\big) = \int_{\sigma - j\omega}^{\sigma + j\omega} Y(s)e^{st}\, ds = y(t), \qquad t \geq 0 \tag{3.20}$$

where σ is a selected value to the right of all the singularities of $Y(s)$ in the s plane.

In practice, this relation is seldom used, as the Laplace transform $Y(s)$ is given as a proper rational function in s

$$Y(s)=\frac{N(s)}{D(s)}=\frac{n_m s^m+n_{m-1}s^{m-1}+\cdots+n_1 s+n_0}{s^n+d_{n-1}s^{n-1}+\cdots+d_1 s+d_0} \tag{3.21}$$

with $m \leq n$, such that this complex Laplace transform will be broken down into simpler ones that are listed in table 3.2 along with their corresponding time responses.

For this reason, we apply the partial fraction expansion method to obtain such simpler transforms, by calculating in first place the roots of $D(s)$, making

$$D(s)=0=(s-p_1)^{r_1}(s-p_2)^{r_2}\ldots(s-p_l)^{r_l} \tag{3.22}$$

for p_i , $i=1\ldots l$ distinct roots, with multiplicity of r_l and complex in general.

By factoring the rational function $Y(s)$ into its partial fraction expansion, we have in case of single roots and distinct

$$Y(s)=\frac{N(s)}{D(s)}=\frac{K_1}{(s+p_1)}+\frac{K_2}{(s+p_2)}+\cdots+\frac{K_l}{(s+p_l)} \tag{3.23}$$

where the constants K_j or residuals of $Y(s)$ are given by

$$K_j=\lim_{s\to -p_j} Y(s)(s+p_j) \qquad j=1\ldots l \tag{3.24}$$

while in case of multiple roots we decompose $Y(s)$ in

$$Y(s)=\frac{N(s)}{D(s)}=\frac{K_{11}}{(s+p_1)}+\frac{K_{12}}{(s+p_1)^2}+\cdots+\frac{K_{1r_1}}{(s+p_1)^{r_1}}++\frac{K_{21}}{(s+p_2)} \tag{3.25}$$

$$+\frac{K_{22}}{(s+p_2)^2}+\cdots+\frac{K_{2r_2}}{(s+p_2)^{r_2}}+\cdots+\frac{K_{l1}}{(s+p_l)}+\frac{K_{l2}}{(s+p_l)^2}+\cdots$$
$$+\frac{K_{lr_l}}{(s+p_l)^{r_l}}$$

where the residuals K_{ji} of $Y(s)$ are defined by

$$K_{ji}=\frac{1}{(r_j-i)!}\lim_{s\to -p_j}\frac{d^{r_j-i}\left(Y(s)(s+p_j)^{r_j}\right)}{ds^{r_j-i}} \qquad j=1\ldots l,\quad i=1,\ldots r_j \tag{3.26}$$

Once the residuals are determined, we proceed to calculate the inverse Laplace transforms of each individual fraction of (3.25) by identifying the corresponding time function associated in table 3.2, and then use the superposition relation to obtain the desired function $y(t)$.

In case of complex conjugate roots, it is necessary to group the corresponding partial fractions prior to identifying the inverse transform in table 3.2. Assuming that the complex roots are located in $p_1 = -\sigma + j\,\omega$ and $p_2 = -\sigma - j\,\omega$, the corresponding partial fractions are given by

$$Y(s) = \frac{N(s)}{D(s)} = \cdots + \frac{K_1}{s + \sigma - j\,\omega} + \frac{K_2}{s + \sigma + j\,\omega} + \cdots$$

Applying (3.24) we obtain the residuals

$$K_1 = \lim_{s \to -\sigma + j\omega} Y(s)(s + \sigma - j\omega)$$

$$K_2 = \lim_{s \to -\sigma - j\omega} Y(s)(s + \sigma + j\omega)$$

where $K_1 = \alpha_1 + j\beta_1$ and $K_2 = \alpha_1 - j\beta_1$ a pair if complex conjugate residuals.

Grouping the two partial fractions we have

$$Y(s) = \frac{N(s)}{D(s)} = \cdots + \frac{\widehat{K}_1(s + \sigma)}{(s + \sigma)^2 + \omega^2} + \frac{\widehat{K}_2\omega}{(s + \sigma)^2 + \omega^2} + \cdots$$

with $\widehat{K}_1 = 2\alpha_1$ and $\widehat{K}_2 = -2\beta_1$, so it is now possible to calculate the inverse Laplace transforms of each individual fraction by identifying the corresponding time function associated in table 3.2.

Example 3.4

Obtain the step response $\mathrm{y(t)}$ corresponding to fig. 3.9 starting from the preceding value of $Y(s)$, using the MATLAB commands necessary to derive the residues and to plot the time response.

Solution

Starting from $Y(s)$ of (3.31), we calculate the roots of the denominator

$$s(s^2 + 6s + 9) = 0$$

by applying the MATLAB code

```
>> den= [1 6 9 0];
>> roots(den);

ans =

         0
   -3.0000
   -3.0000
```

In this way we can rewrite $Y(s)$ as the partial fraction expansion

$$Y(s) = \frac{1}{s(s^2+6s+9)} = \frac{K_1}{s} + \frac{K_{21}}{(s+3)} + \frac{K_{22}}{(s+3)^2}$$

Using (3.24) and (3.26) we obtain the residues

$$K_1 = \lim_{s\to 0} Y(s)\, s \ = \frac{1}{9}$$

$$K_{21} = \frac{1}{2-1!}\lim_{s\to -3} \frac{d(Y(s)(s+3)^2)}{ds} = \lim_{s\to -3} \frac{d}{ds}(\frac{1}{s}) = \lim_{s\to -3} -\frac{1}{s^2} = -\frac{1}{9}$$

$$K_{22} = \frac{1}{2-2!}\lim_{s\to -3} Y(s)(s+3)^2 = \lim_{s\to -3} \frac{1}{s} = -\frac{1}{3}$$

We can check the results obtained by applying the MATLAB code

```
>> num=[1];den=[1 6 9 0];
>> [r,p]=residue(num,den)

r =

   -0.1111
   -0.3333
    0.1111

p =

    -3
    -3
     0
```

In this way, the partial fraction expansion becomes

$$Y(s) = \frac{1}{s(s^2+6s+9)} = \frac{1}{9s} - \frac{1}{9(s+3)} - \frac{1}{3(s+3)^2}$$

so that the time response can be calculated adding the corresponding inverse transform of each fraction in (3.47), yielding the solution

$$y(t) = \frac{1}{9}(1 - e^{-3t} - 3te^{-3t})u_s(t)$$

In fig. 3.10 it is shown that the step response after applying the MATLAB code

```
>> t = 0:0.1:5; y= 0.1111*(1-exp(-3*t)-3*t.*exp(-3*t));
>> plot(t,y)
```

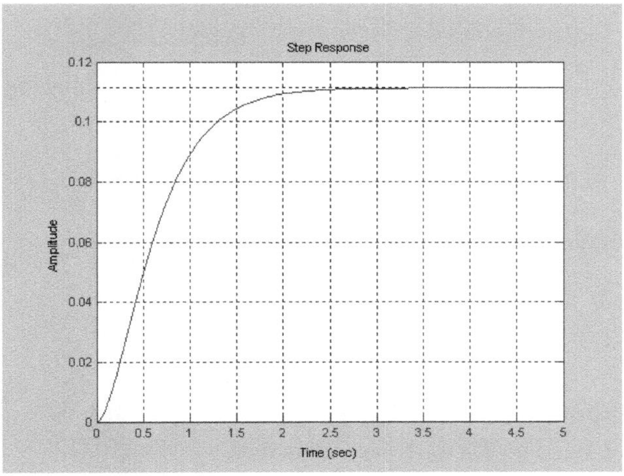

Fig. 3.10 Step response of the mechanical system of example 3.3

Example 3.5

Obtain the time response $v_0(t)$ to input signal $i_S(t) = 0.5u_s(t)$ for the electric circuit shown in fig. 3.11 whose dynamic equation is given by

$$LC\frac{d^2v_0}{dt^2} + (R_1 + R_2)C\frac{dv_0}{dt} + v_o = R_1 i_s$$

for values of $R_1 = 0.2$, $R_2 = 1$, $C = 0.01$ and $L = 0.1$, starting from initial conditions $v_0(0) = 0$ and $\frac{dv_0}{dt}(0) = 0$.

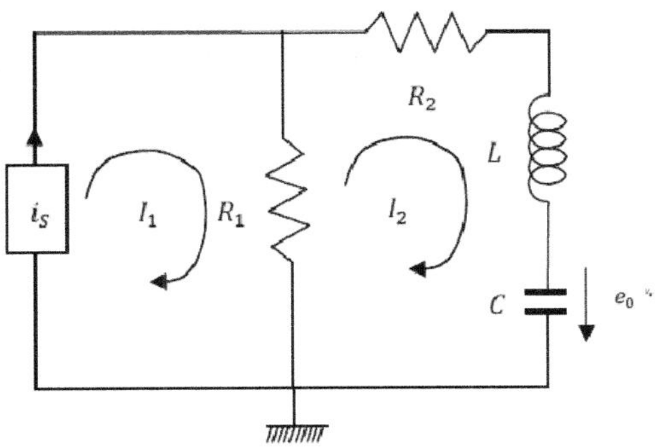

Fig. 3.11 Diagram of the electric circuit for example 3.5

Solution:

Applying the Laplace transform properties to the differential equation of the circtuit we get

$$LCs^2\,V_0(s) + (R_1 + R_2)Cs\,V_0(s) + V_0(s) = R_1 I_S(s)$$

so that the system response is given by

$$V_0(s) = \frac{R_1}{LCs^2 + (R_1 + R_2)Cs + 1}\,I_S(s)$$

In case of step input $i_S(t) = 0.5u_s(t)$, $I_s(s) = \frac{0.5}{s}$, so that the system response $V_0(s)$ particularized for R_1, R_2, C and L is given by

$$V_0(s) = \frac{0.2\;0.5}{s(0.001s^2 + 0.012s + 1)}$$

Now we proceed to calculate the roots of the denominator

$$s(0.001s^2 + 0.012s + 1) = 0$$

by applying the MATLAB code

```
>> den= [0.001 0.012 1 0];
>> roots(den)

ans =

         0
  -6.0000 +31.05i
  -6.0000 -31.05i
```

Expanding $V_0(s)$ into partial fractions, we have

$$V_0(s) = \frac{0.1}{s(0.001s^2 + 0.012s + 1)} = \frac{100}{s(s^2 + 12 + 1000)}$$

$$V_0(s) = \frac{100}{s(s^2 + 12 + 1000)} = \frac{K_1}{s} + \frac{K_2}{s + 6 - j\,31.05} + \frac{K_3}{s + 6 + j\,31.05}$$

Using (3.24) we obtain the residues

$$K_1 = \lim_{s\to 0} V_0(s)\,s = \frac{100}{1000} = 0.1$$

$$K_2 = \lim_{s \to -6+j\,31.05} V_0(s)(s+6-j\,31.05) = \lim_{s \to -6+j\,31.05} \frac{100}{s(s+6+j\,31.05)} = -0.05 + j\,0.0097$$

$$K_3 = \lim_{s \to -6-j\,31.05} V_0(s)(s+6+j\,31.05) = \lim_{s \to -6-j\,31.05} \frac{100}{s(s+6-j\,31.05)} = -0.05 - j\,0.0097$$

We can check the results obtained by applying the MATLAB code

```
>> num=100;den=[1 12 1000 0];
>> [r,p]=residue(num,den)

r =

  -0.0500 + 0.0097i
  -0.0500 - 0.0097i
   0.1000


p =

  -6.0000 +31.05i
  -6.0000 -31.05i
        0
```

In this way, the partial fraction expansion becomes

$$V_0(s) = \frac{100}{s(s^2+12+1000)} = \frac{0.1}{s} + \frac{-0.05+j\,0.0097}{s+6-j\,31.05}$$

$$\frac{-0.05-j\,0.0097}{s+6+j\,31.05} +$$

By grouping the two complex fractions we have

$$V_0(s) = \frac{0.1}{s} - \frac{0.1(s+6)}{(s+6)^2+(31.05)^2} - \frac{0.0194 \cdot 31.05}{(s+6)^2+(31.05)^2}$$

so that the time response can be calculated adding the corresponding inverse transform of each fraction, yielding the solution

$$v_0(t) = (0.1 - 0.1e^{-6t}\cos(31.05\,t) - 0.6e^{-6t}\sin(31.05\,t))u_s$$

In fig. 3.12 the step response after applying the MATLAB code is shown

```
>> t = 0:0.2:2; y= 0.1*(1-exp(-6*t).cos(31.05*t)-6*exp(-6*t).*sin(31.05*t));
>> plot(t,y)
```

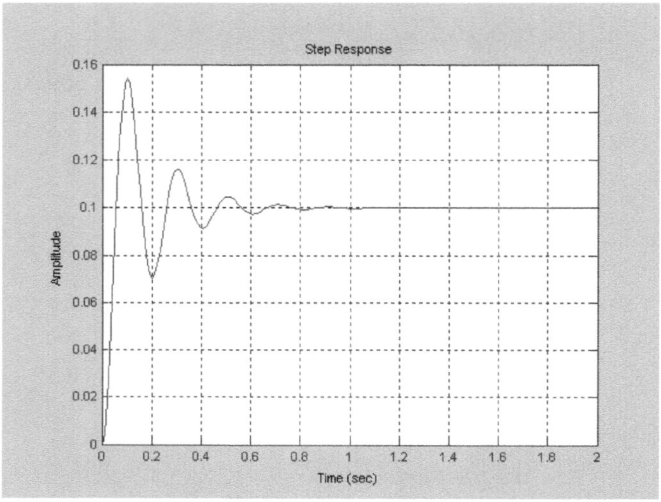

Fig. 3.12 Step response of the electric circuit of example 3.5

3.5 Transfer Function

The transfer function of an invariant linear system is defined as the ratio between the Laplace transform of system's output $Y(s)$ and the transform of the system's input $U(s)$ under the assumption of zero initial conditions. This function enables the separation of the input, system, and output into three separate and distinct parts, unlike the differential equation which define its dynamics.

Considering the single-input single-output continuous-time system defined by a linear and invariant differential equation in the form

$$a_n y^{n)} + \cdots + a_1 \dot{y} + a_0 y = b_m\, u^{m)} + \cdots + b_1 \dot{u} + b_0 u \tag{3.27}$$

and taking the Laplace transform of both sides, we have

$$\begin{aligned} a_n s^n Y(s) + \cdots + a_1 s Y(s) + a_0 Y(s) &= b_m s^m U(s) + \cdots \\ &+ b_1 s U(s) + b_0 U(s) \end{aligned} \tag{3.28}$$

by assuming that the initial conditions are null. Taking the ratio between $Y(s)$ and $U(s)$, we obtain the transfer function of this system as defined by

$$G(s) = \frac{Y(s)}{U(s)} = \frac{b_m s^m + \cdots + b_1 s + b_0}{a_n s^n + \cdots + a_1 s + a_0} \tag{3.29}$$

The transfer function is a property of the system itself, since it is not depend on the input, and can be represented as a block diagram (fig. 3.13).

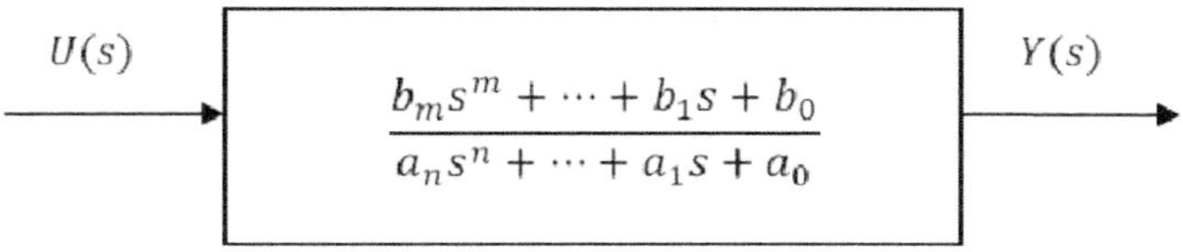

Fig. 3.13 Block diagram of the transfer function $G(s)$.

In fact, the output response $Y(s)$ for an input signal $U(s)$ can be obtained by using the relation

$$Y(s) = G(s)U(s) \tag{3.30}$$

assuming there are no initial conditions.

The highest power of the denominator of $G(s)$ (characteristic equation) is called the system order. The roots of the characteristic equation are called the system poles, while the roots of the numerator are called the system zeros (fig. 3.14)

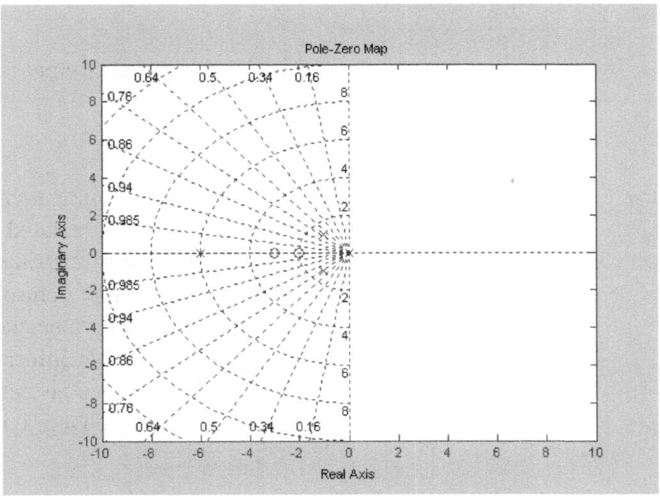

Fig. 3.14 Pole-zero diagram in the s plane for the transfer function $G(s) = \frac{(s+3)(s+2)}{s(s+1+j)(s+1-j)(s+6)}$.

This form of representation corresponds to the external description of the system, which provides no information on its internal structure. Moreover, the transfer function of different systems may be the same, which is the case of analog systems.

Example 3.6

Obtain the transfer function of the hydraulic system of fig. 3.15 by considering $P_s(t)$ and $P_{3r}(t)$ as input and output signal respectively, with a dynamic equation given by

$$CI\frac{d^2P_{3r}}{dt^2} + RC\frac{dP_{3r}}{dt} + P_{3r} = P_s$$

and indicate the poles and zeros locations inside the s plane for $C = 2$, $R = 3$, and $I = 0.5$. Then, derive and plot the step response using the Laplace transform. Create the transfer function and obtain the step response by using the MATLAB appropriate commands, comparing both responses.

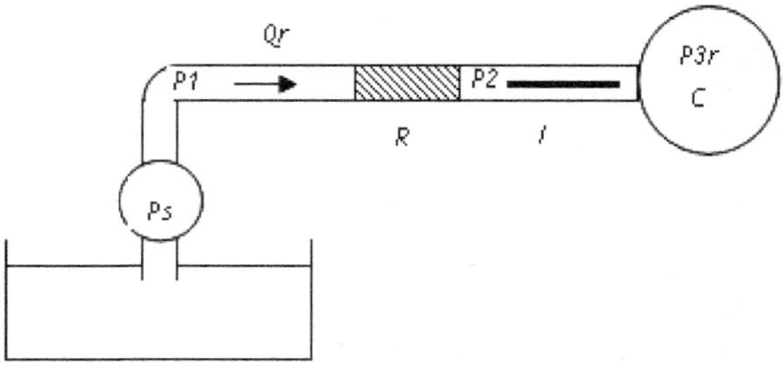

Fig. 3.15 Hydraulics system of example 3.6

Solution

Applying the Laplace transform, the differential equation assuming zero initial conditions we have

$$CIs^2P_{3r}(s) + RCsP_{3r}(s) + P_{3r}(s) = P_s(s)$$

Then, taking the ratio between $P_{3r}(s)$ and $P_s(s)$ we obtain the transfer function

$$G(s) = \frac{P_{3r}(s)}{P_s(s)} = \frac{1}{CIs^2 + RCs + 1}$$

Substituting for C, R and I we have

$$G(s) = \frac{P_{3r}(s)}{P_s(s)} = \frac{1}{s^2 + 6s + 1}$$

The system's poles are located in the solutions of the characteristic equation

$$s^2 + 6s + 1 = 0$$

For this calculation we use the MATLAB command

```
>> roots([1 6 1])

ans =

   -5.8284
   -0.1716
```

while there are no zeros for this system.

In order to calculate the step response, we obtain $P_s(s) = \mathcal{L}(P_s(t)) = \frac{1}{s}$, so that

$$P_{3r}(s) = \frac{1}{s^2 + 6s + 1} P_s(s) = \frac{1}{s(s^2 + 6s + 1)}$$

The partial fraction expansion becomes

$$P_{3r}(s) = \frac{1}{s(s^2 + 6s + 1)} = \frac{K_1}{s} + \frac{K_2}{(s + 0.1716)} + \frac{K_3}{(s + 5.8284)}$$

and the residuals are obtained using the MATLAB code

```
>> num=[1];den=[1 6 1 0];
>> [r,p]=residue(num,den)

r =

    0.0303
   -1.0303
    1.0000

p =

   -5.8284
   -0.1716
         0
```

In this way, the partial fraction expansion becomes

$$P_{3r}(s) = \frac{1}{s} - \frac{1.0303}{(s + 0.1716)} + \frac{0.0303}{(s + 5.8284)}$$

so that the time response can be calculated adding the corresponding inverse transform of each fraction in (3.47), yielding the solution

$$P_{3r}(t) = (1 - 1.0303e^{-0.1716t} + 0.0303e^{-5.8284t})u_s(t)$$

In fig. 3.16 the step response after applying the MATLAB code is shown

```
>> t = 0:0.1:5; y= (1-1.0303*exp(-0.1716*t)+0.0303*exp(-5.8284*t));
>> plot(t,y)
```

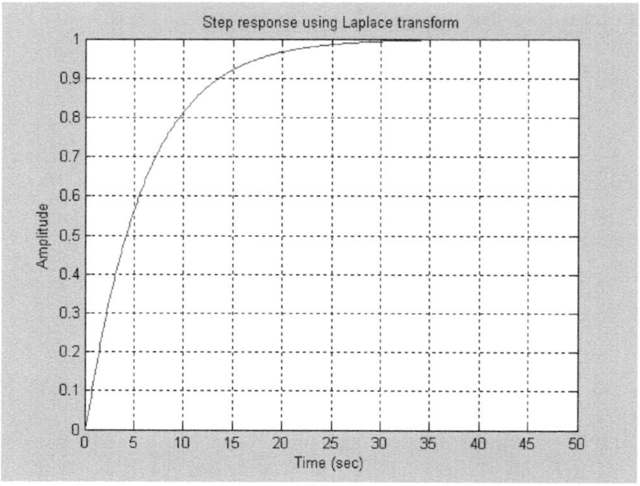

Fig. 3.16 Step response of the hydraulic system of example 3.6

To create the transfer function directly we use the following MATLAB code

```
>> s=tf('s');
>> G=1/(s^2+6*s+1)

Transfer function:
      1
-------------
s^2 + 6 s + 1

>> step(G,50)
```

In fig. 3.17 the step response of the system running under MATLAB is shown.

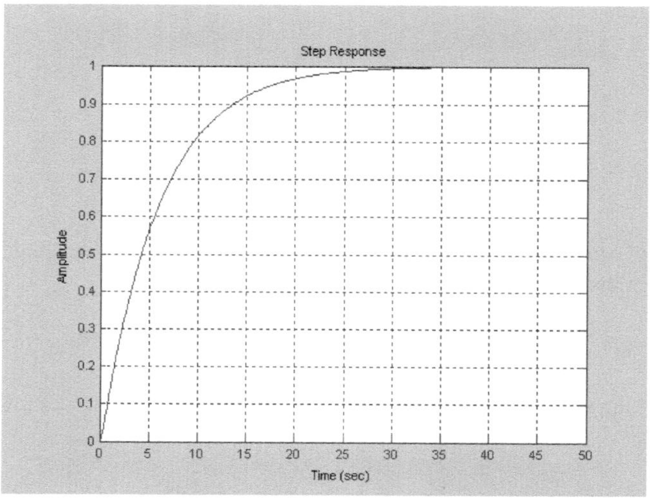

Fig. 3.17 MATLAB step response of the hydraulic system of example 3.6.

In the case of multi-input multi-output systems, we define the transfer matrix $\bar{G}(s)$, whose elements $G_{ij}(s)$ relate each output $Y_i(s)$ to each input $U_j(s)$, assuming zero both the remaining entries $U_k(s), k \neq j$ and the initial conditions (fig. 3.18).

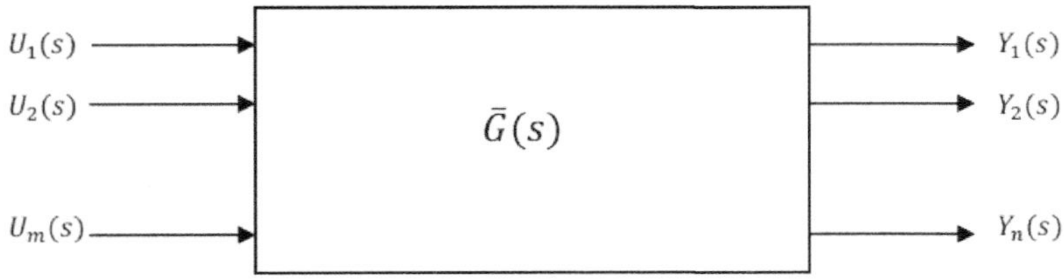

Fig. 3.18 Continuous multi-input, multi-output system.

Each individual transfer function element $G_{ij}(s)$ is given by the relation

$$G_{ij}(s) = \frac{Y_i(s)}{U_j(s)} \tag{3.31}$$

and assuming no initial conditions, we derive each of the system output, following the expressions

$$\begin{aligned} Y_1(s) &= G_{11}(s)U_1(s) + \cdots + G_{1m}(s)U_m(s) \\ Y_2(s) &= G_{21}(s)U_1(s) + \cdots + G_{2m}(s)U_m(s) \\ &\vdots \\ Y_n(s) &= G_{n1}(s)U_1(s) + \cdots + G_{nm}(s)U_m(s) \end{aligned} \tag{3.32}$$

which can be transformed to a vector notation

$$\bar{Y}(s) = \bar{G}(s)\bar{U}(s) \tag{3.33}$$

assuming there are no initial conditions, with $\bar{Y}(s) = [Y_1(s) \dots Y_n(s)]$ and $\bar{U}(s) = [U_1(s) \dots U_m(s)]$.

3.6 The State Space Approach

The state of a dynamic system is defined by the minimum set of variables $\bar{x}(t) = \{x_1(t), x_2(t), \dots, x_n(t)\}$ such that the knowledge of the value of $\bar{x}(t_0)$ together with the knowledge of the input $\bar{u}(t), t \geq t_0$, completely determines the system behavior for $t \geq t_0$. These variables are usually well-known and measurable quantities.

The set of state variables constitutes a vector, so that the system behavior can be described in a state space with n axis corresponding to each state vector's component $x_i(t)$. Assuming that the actual system state at any time t in the state space is represented by a point, we can define a state trajectory as the path over time of these points as the system evolves for $t \geq t_0$.

The state space representation (internal description) is useful for analyzing multi-input multi-output systems, and is given in general by the state transition and the state output equations. In general, considering a continuous system with r inputs and s outputs (fig. 3.19) the state transition equations can be described by a set of n independent differential equations in the following form

$$\begin{aligned} \dot{x}_1 &= \emptyset_1(x_1, x_2, \dots, x_n, u_1, u_2, \dots, u_r, t) \\ \dot{x}_2 &= \emptyset_2(x_1, x_2, \dots, x_n, u_1, u_2, \dots, u_r, t) \\ &\vdots \\ \dot{x}_n &= \emptyset_n(x_1, x_2, \dots, x_n, u_1, u_2, \dots, u_r, t) \end{aligned} \tag{3.34}$$

with $\emptyset_i, i = 1, \dots, n$ non-linear in general. The parameter n is denoted as the system order, and plays a fundamental role for the system behavior.

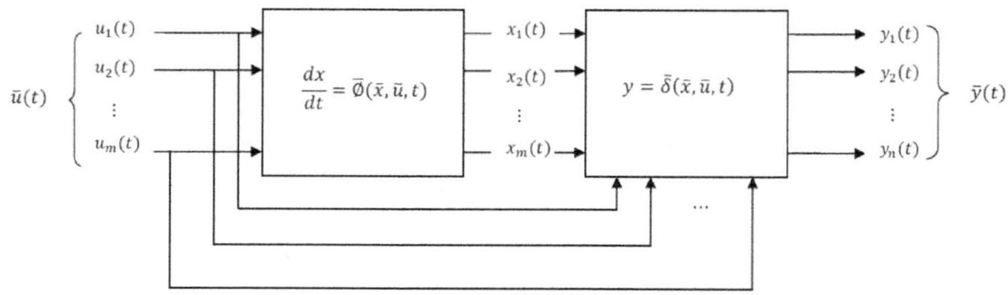

Fig. 3.19 State space representation for a continuous system

In the same way, the state output equations can be described by a set of s independent differential equations in the following form

$$\begin{aligned} y_1 &= \delta_1(x_1, x_2, \dots, x_n, u_1, u_2, \dots, u_r, t) \\ y_2 &= \delta_2(x_1, x_2, \dots, x_n, u_1, u_2, \dots, u_r, t) \\ &\vdots \\ y_s &= \delta_s(x_1, x_2, \dots, x_n, u_1, u_2, \dots, u_r, t) \end{aligned} \tag{3.35}$$

also with $\delta_i, i = 1, \dots, s$ non-linear in general.

Let us particularize the analysis for a dynamical system with $\emptyset_i, i = 1, \dots, n$ and $\delta_j, j = 1, \dots, s$ both described by a linear combination of state and input signals, assuming also the system is stationary with no explicit direct dependence with t. In this case, the state transition equations take the form

$$\begin{aligned} \dot{x}_1 &= a_{11}x_1 + \cdots + a_{1n}x_n + b_{11}u_1 + \cdots + b_{1r}u_r \\ \dot{x}_2 &= a_{21}x_1 + \cdots + a_{2n}x_n + b_{21}u_1 + \cdots + b_{2r}u_r \\ &\vdots \\ \dot{x}_n &= a_{n1}x_1 + \cdots + a_{nn}x_n + b_{n1}u_1 + \cdots + b_{nr}u_r \end{aligned} \tag{3.36}$$

while the state output equations are represented by

$$\begin{aligned} y_1 &= c_{11}x_1 + \cdots + c_{1n}x_n + d_{11}u_1 + \cdots + d_{1r}u_r \\ y_2 &= c_{21}x_1 + \cdots + c_{2n}x_n + d_{21}u_1 + \cdots + d_{2r}u_r \\ &\vdots \\ y_s &= c_{s1}x_1 + \cdots + c_{sn}x_n + d_{s1}u_1 + \cdots + d_{sr}u_r \end{aligned} \tag{3.37}$$

The state equations (3.36) and (3.37) can be rewritten by using the vector-matrix representation, hence

$$\begin{aligned} \frac{d\bar{x}}{dt} &= A\bar{x}(t) + B\bar{u}(t) \\ \bar{y}(t) &= C\bar{x}(t) + D\bar{u}(t) \end{aligned} \tag{3.38}$$

where A is a $n \times n$ matrix, B is a $n \times r$ matrix, C is a $s \times n$ matrix and D is a $s \times r$ matrix, constituting $\{A, B, C, D\}$, the state space representation of a linear and invariant dynamic system. This state space representation is not unique, since each one depends on the state variable selection procedure. In general, the set of state variable selected includes one or more variables and its successive derivatives.

Example 3.7

Derive the state equations for the electromechanical system shown in fig. 3.20 starting from the dynamic equations written as

$$L_a\frac{di_a}{dt} + R_a i_a(t) + e_m(t) = e_a(t)$$

$$e_m(t) = K_m\omega(t)$$

$$\tau(t) = K_m i_a(t)$$

$$J\frac{d\omega}{dt} + B\omega(t) = \tau(t)$$

by using a state vector $\bar{x}(t) = \{x_1(t), x_2(t)\} = \{i_a(t), \omega(t)\}$, where $u(t) = e_a(t)$ and $y(t) = \omega(t)$. Represent this system by using MATLAB and obtain the response for an input signal $e_a(t) = exp(-t)\,sin\,(2t)$, for the parameter values of $L_a = 1$, $R_a = 5$, $K_m = 0.8$, $J = 1.5$ and $B = 0.2$.

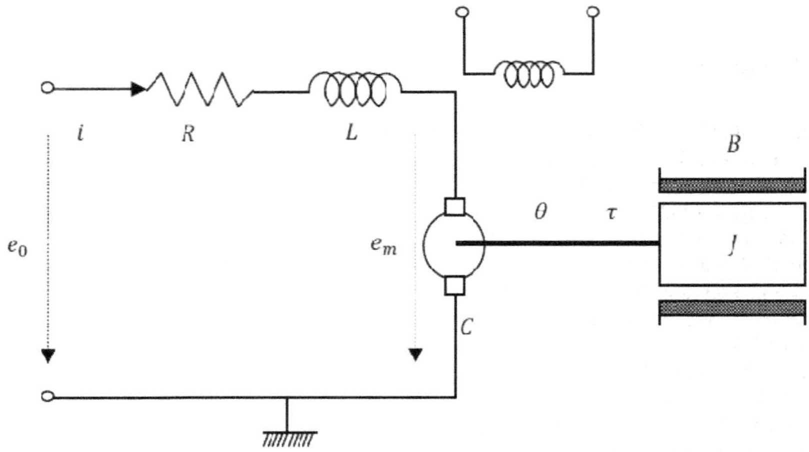

Fig. 3.20. Diagram of a DC motor

Solution

In order to derive the state transition equations we solve for the derivatives of $\bar{x}(t) = \{i_a(t), \omega(t)\}$. Therefore, we substitute $e_m(t)$ and $\tau(t)$ into the other equations, so we have

$$L_a \frac{di_a}{dt} = e_a - R_a i_a - K_m \omega$$

$$J \frac{d\omega}{dt} = K_m i_a - B\omega(t)$$

In this way, solving for the derivatives of $\bar{x}(t)$ and using the corresponding assignment we get

$$\frac{dx_1}{dt} = \frac{1}{L_a} u - \frac{R_a}{L_a} x_1 - \frac{K_m}{L_a} x_2 = \emptyset_1(x_1, x_2, u)$$

$$\frac{dx_2}{dt} = \frac{K_m}{J} x_1 - \frac{B}{J} x_2 = \emptyset_2(x_1, x_2, u)$$

Since the system output has been already assigned as $y(t) = \omega(t)$, the state output equation becomes

$$y = x_2 = \delta(x_1, x_2, u)$$

Rewriting in vector-matrix form yields

$$\begin{bmatrix} \frac{di_a}{dt} \\ \frac{d\omega}{dt} \end{bmatrix} = \begin{bmatrix} -\frac{R_a}{L_a} & -\frac{K_m}{L_a} \\ \frac{K_m}{J} & -\frac{B}{J} \end{bmatrix} \begin{bmatrix} i_a \\ \omega \end{bmatrix} + \begin{bmatrix} B \\ 0 \end{bmatrix} u$$

$$y = \begin{bmatrix} 0 & 1 \end{bmatrix} \begin{bmatrix} i_a \\ \omega \end{bmatrix}$$

Encoding the system dynamics into MATLAB we get

```
>> La=1;Ra=5;Km=0.8;J=1.5;B=0.2;
>> A_m=[-Ra/La -Km/La;Km/J -B/J]

A_m =

   -5.0000   -0.8000
    0.5333   -0.1333

>> B_m=[B;0]

B_m =

    0.2000
         0

>> C_m=[0 1]

C_m =

0    1

>>D_m=[0]
>> SYS=ss(A_m,B_m,C_m,D_m)

a =
                x1        x2
   x1           -5      -0.8
   x2       0.5333   -0.1333

b =
          u1
   x1    0.2
   x2      0

c =
          x1   x2
   y1      0    1

d =
          u1
   y1      0

Continuous-time model.
>> t=0:0.1:20;u=10*exp(-t).*sin(2*t)
>> y=lsim(SYS,u,t);plot(t,y)
>> grid
```

In fig. 3.21 the response of the DC-motor running under MATLAB is shown.

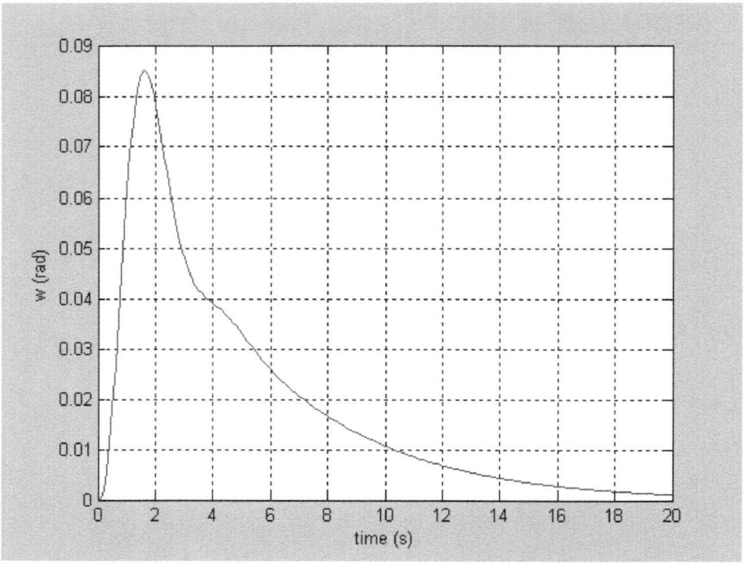

Fig. 3.21 Response of the DC motor to input signal $e_a(t)$

3.6.1 Relation between Transfer Function and State Space Approaches

Both the transfer function approach and the state equation formulation represent the dynamics of the same continuous linear and invariant system. Therefore, it is obvious to expect that there is a interrelation between both kind of representation.

In effect, we are going to consider a simple case of single-input single-output linear and invariant system similar to the one described in (3.3) defined by

$$a_n y^{n)} + \cdots + a_1 \dot{y} + a_0 y = b_n\, u^{n)} + \cdots + b_1 \dot{u} + b_0 u \tag{3.39}$$

with $a_n = 1$ (otherwise we divide both members of (3.39) by a_n) and $b_i, i = 0, \dots n$ (in general $b_j = 0$ for $i = m+1, \dots n$ if $m \le n$).

We are going to define the set of state variables as

$$\begin{aligned} x_1 &= y - \beta_0 u \\ x_2 &= \dot{y} - \beta_o \dot{u} - \beta_1 u \\ &\vdots \\ x_n &= y^{n-1)} - \beta_0 u^{n-1)} - \cdots - \beta_{n-1} u \end{aligned} \tag{3.40}$$

where the parameters $\beta_j, j = 0, \dots n-1$ are given by

$$\begin{aligned}\beta_0 &= b_n \\ \beta_1 &= b_{n-1} - a_{n-1}\beta_0 \\ &\vdots \\ \beta_n &= b_0 - a_{n-1}\beta_{n-1} - \cdots - a_1\beta_1 - a_0\beta_0\end{aligned} \tag{3.41}$$

Then, according to (3.40) definition, the state transition equation becomes

$$\begin{aligned}\dot{x}_1 &= x_2 + \beta_1 u \\ \dot{x}_2 &= x_3 + \beta_2 u \\ &\vdots \\ \dot{x}_n &= -a_0 x_1 + \cdots - a_{n-1}x_n + \beta_n\end{aligned} \tag{3.42}$$

Rewriting (3.42) in vector-matrix form we have

$$\begin{bmatrix}\dot{x}_1 \\ \dot{x}_2 \\ \vdots \\ \dot{x}_n\end{bmatrix} = \frac{d\bar{x}}{dt} = \begin{bmatrix}0 & 1 & \dots & 0 \\ 0 & 0 & \dots & 0 \\ \vdots & \vdots & \vdots & \vdots \\ -a_0 & -a_1 & \cdots & -a_{n-1}\end{bmatrix}\begin{bmatrix}x_1 \\ x_2 \\ \vdots \\ x_n\end{bmatrix} + \begin{bmatrix}\beta_1 \\ \beta_2 \\ \vdots \\ \beta_n\end{bmatrix} u = A\bar{x} + Bu \tag{3.43}$$

Recalling that the output is defined by (3.40) we get

$$y = x_1 + \beta_0 u \tag{3.44}$$

and rewriting (3.44) in vector-matrix form we have

$$y = [1 \quad 0 \quad \dots \quad 0]\begin{bmatrix}x_1 \\ x_2 \\ \vdots \\ x_n\end{bmatrix} + [\beta_0]u = C\bar{x} + Du \tag{3.45}$$

Therefore, it is possible to transform the transfer function representation into the state space one, simply by identifying the corresponding parameters $a_j, j = 0, \dots n$ and $b_k, k = 0, \dots n$ of the transfer function corresponding to (3.39), to obtain after the state space matrices $\{A, B. C, D\}$.

Conversely, the transfer function can be easily determined from the state space representation $\{A, B. C, D\}$ by simply applying the Laplace transform to the definition equations as described in (3.38) (assuming null initial conditions). Then, we have

$$\begin{aligned}s\bar{X}(s) &= A\bar{X}(s) + B\bar{U}(s) \\ \bar{Y}(s) &= C\bar{X}(s) + D\bar{U}(s)\end{aligned} \tag{3.46}$$

Rearranging the $\bar{X}(s)$ term we find that

$$(sI - A)\bar{X}(s) = B\bar{U}(s) \tag{3.47}$$

$$\bar{X}(s) = (sI - A)^{-1}B\bar{U}(s) \tag{3.48}$$

and substituting, we obtain

$$\bar{Y}(s) = (C(sI - A)^{-1}B + D)\bar{U}(s) \tag{3.49}$$

which defines the transfer matrix function (in general) of the given system

$$\bar{G}(s) = C(sI - A)^{-1}B + D \tag{3.50}$$

The results obtained here can be easily extended to multiple-input multiple-output systems increasing the size of the matrices B, C, and D according to the dimension of the input and output signals $\bar{u}(t)$ and $\bar{y}(t)$.

Example 3.8

Derive the state space equations corresponding to the hydraulic system of Example 3.6 with transfer function defined by

$$G(s) = \frac{P_{3r}(s)}{P_s(s)} = \frac{1}{s^2 + 6s + 1}$$

and realize the transfer function to state space conversion by using adequate MATLAB commands and plot also the step response.

Solution

Starting from the transfer function we identify by inspection the coefficients $a_j, j = 0, \ldots n$ and $b_k, k = 0, \ldots n$ with $n = 2$, so we have $a_0 = 1, a_1 = 6, a_2 = 1$ while $b_0 = 1,\ b_1 = 0, b_2 = 0$. Therefore we get

$$\beta_0 = b_2 = 0$$

$$\beta_1 = b_1 - a_1\beta_0 = 0$$

$$\beta_2 = b_0 - a_1\beta_1 - a_0\beta_0 = 1$$

By substituting in (3.43) we have the state equations

$$\begin{bmatrix}\dot{x}_1\\ \dot{x}_2\end{bmatrix} = \begin{bmatrix}0 & 1\\ -1 & -6\end{bmatrix}\begin{bmatrix}x_1\\ x_2\end{bmatrix} + \begin{bmatrix}0\\ 1\end{bmatrix}u = A\bar{x} + Bu$$

$$y = \begin{bmatrix}1 & 0\end{bmatrix}\begin{bmatrix}x_1\\ x_2\end{bmatrix} + [0]u = C\bar{x} + Du$$

Encoding the system transfer function into MATLAB and converting it into its state space form we have

```
>> s=tf('s');
>> G=1/(s^2+6*s+1)

Transfer function:
       1
-------------
s^2 + 6 s + 1

>> [A,B,C,D]=ssdata(G)

A =

    -6    -1
     1     0
B =

     1
     0
C =

     0     1
D =

     0

>> ssG=ss(A,B,C,D)



a =
          x1   x2
     x1   -6   -1
     x2    1    0

b =
          u1
     x1    1
     x2    0

c =
          x1   x2
     y1    0    1

d =
          u1
     y1    0

Continuous-time model.
>> step(ssG,50)
```

In fig. 3.22 the step response of the hydraulic system running under MATLAB is shown. It is important to highlight that the state space representation used by MATLAB coincides with the one here described except in the order of the state variables x_1 and x_2 defined.

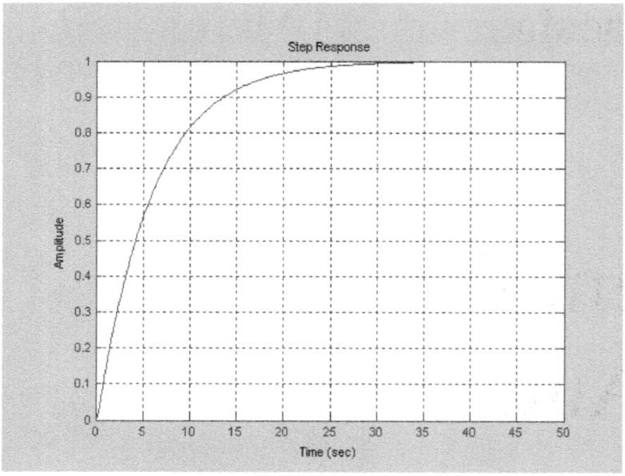

Fig. 3.22 MATLAB step response of the hydraulic system of example 3.8

3.7 Block Diagrams

In many dynamic systems, the system equations are described by the relationship between input and output variables for each one of the component which constitutes the whole system. In these cases it is easy to draw a block diagram that represents the system dynamics and the components that comprises it.

A block diagram is an interconnection of blocks representing basic mathematical operations in such a way that the whole diagram is equivalent to the dynamic equations with its constitutive components. In a block diagram, system variables are represented by lines interconnecting the blocks, and each block represents the mathematical operation that produces the block output on the input signal, expressed as a transfer function.

In addition to the functional blocks, the summing junction also appear to express the addition and subtraction of variables and the pickoff points to transmit system variables to various sections of the block diagram (fig. 3.23)

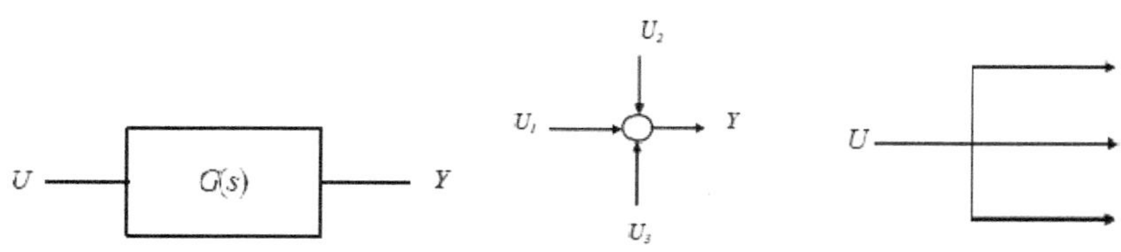

Fig. 3.23 Constitutive elements of a block diagram, block, summing junction and pickoff point

The block diagram is obtained from the equations dynamics that describe the behavior of each component, to which the Laplace transform was previously applied, finally connecting the components of the complete block diagram.

Example 3.9

Obtain the block diagram of the electromechanical system of Example 3.7 starting from the dynamic equations written as

$$L_a \frac{di_a}{dt} + R_a i_a(t) + e_m(t) = e_a(t)$$

$$e_m(t) = K_m \frac{d\theta}{dt}$$

$$\tau(t) = K_m i_a(t)$$

$$J \frac{d^2\theta}{dt^2} + B \frac{d\theta}{dt} = \tau(t)$$

by defining $\theta(t)$ and $e_a(t)$ as the system output and input respectively. Represent the block diagram by using SIMULINK and obtain the step response for the parameter values of $L_a = 1$, $R_a = 5$, $K_m = 0.8$, $J = 1.5$ and $B = 0.2$.

Solution

Applying the Laplace transform to the dynamic equations we have

$$L_a s I_a(s) + R_a I_a(s) + E_m(s) = E_a(s)$$

$$E_m(s) = K_m s\theta(s)$$

$$T(s) = K_m I_a(s)$$

$$Js^2\theta(s) + Bs\theta(s) = T(s)$$

Solving for $I_a(s)$ we get

$$I_a(s) = \frac{1}{L_a s + R_a}(E_a(s) - E_m(s))$$

and solving for $\theta(s)$ we obtain

$$\theta(s) = \frac{1}{Js^2 + Bs} T(s)$$

Based on the relationship between these equations, we can draw the block diagram of the system with input $E_a(s)$ and output θ(s) (fig. 3.24)

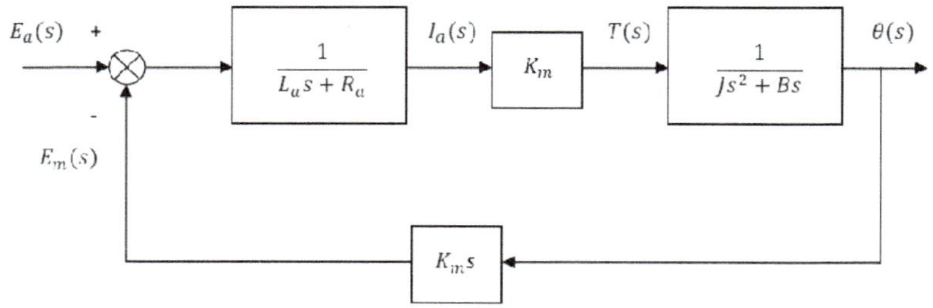

Fig. 3.24 Block diagram of the electromechanical system for Example 3.9.

In fig. 3.25 and fig. 3.26 the corresponding block diagram developed by using SIMULINK together and its step response respectively for the parameter values indicated $L_a = 1$, $R_a = 5$, $K_m = 0.8$, $J = 1.5$ and $B = 0.2$ are shown.

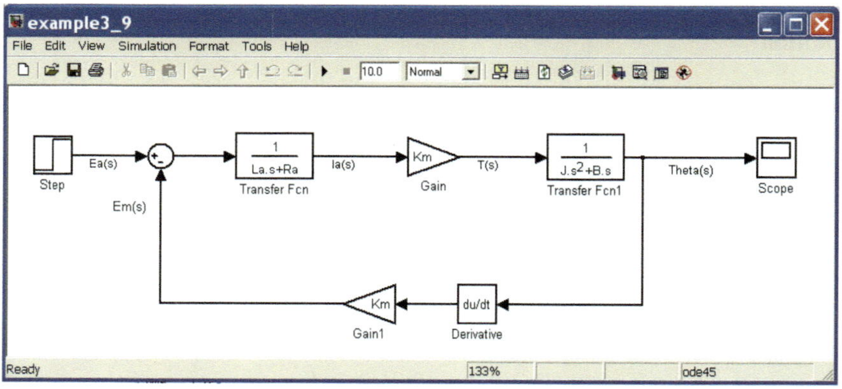

Fig. 3.25 SIMULINK diagram corresponding to Example 3.9

From the block diagram of a system we can make changes to simplify or reduce the original diagram to a single matrix transfer function, or a single transfer function in case of a single-input single-output system, so that we can apply the Laplace transform technique in order to find the system time response.

For this purpose we are going to apply some reduction rules, to transform a more complicated system into a single equivalent system. In table 3.3 the rules commonly used to simplify the block diagrams of dynamic systems are shown.

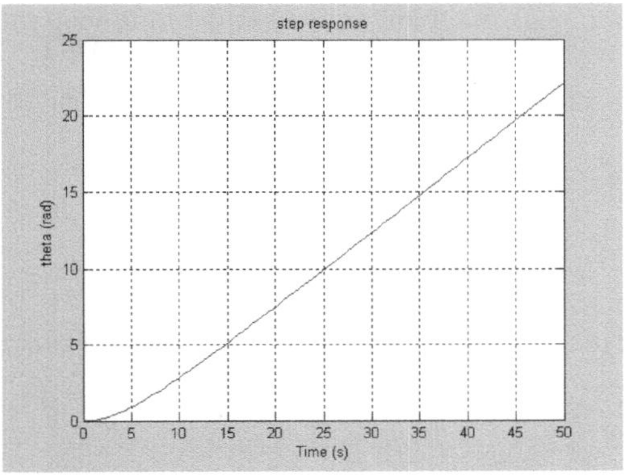

Fig. 3.26 Step response of the DC motor obtained by using SIMULINK

Table 3.3 List of rules commonly used to reduce a block diagram

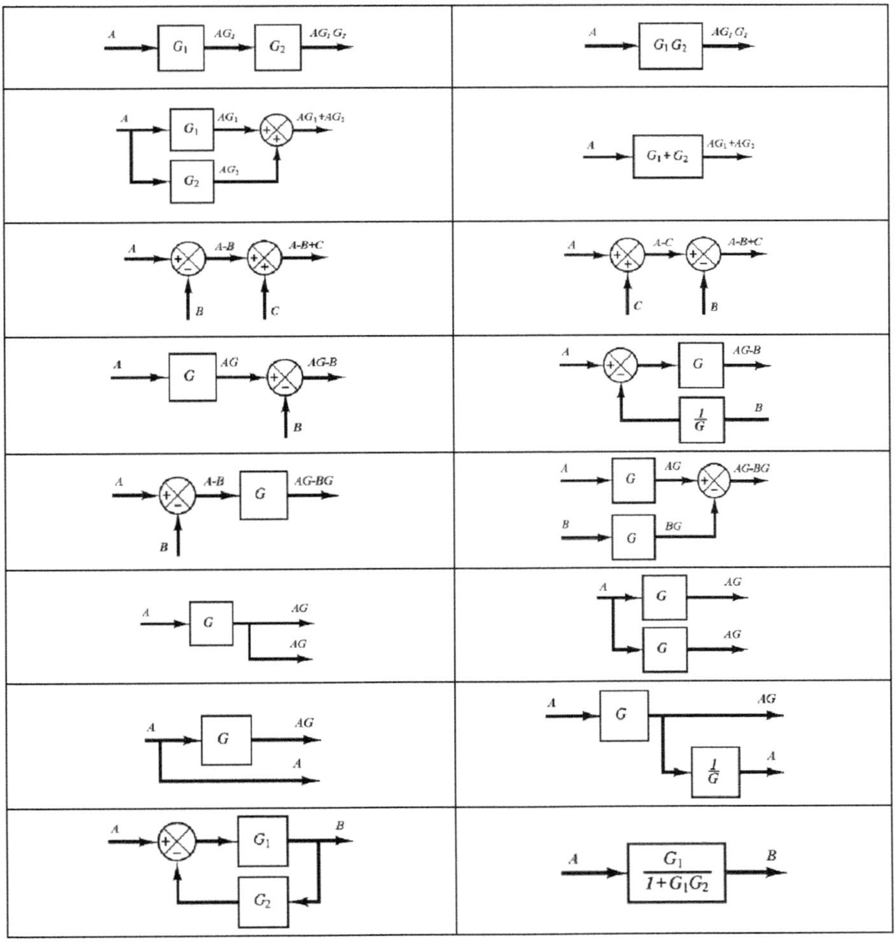

By simplifying a block diagram, remember the following rules:

1. The product of transfer functions in the direction of the direct path must be the same.
2. The product of transfer functions around the loop must be the same.

Example 3.10

Reduce the block diagram of fig. 3.27 to a single transfer function by applying the adequate reduction rules.

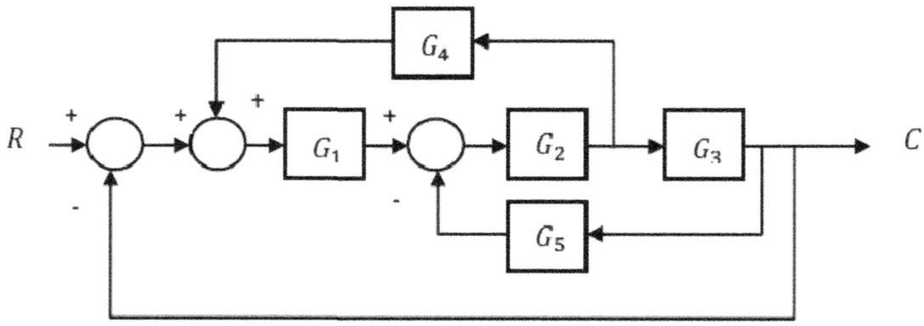

Fig. 3.27 Block diagram for example 3.10

Solution

In first place we move the summing junction between G_1 and G_2 to the left, then solve for the feedback involving G_1, G_2, and G_4. We follow by reducing the feedback implying G_5/G_1 and finally solve for the unity feedback to get the single transfer function demanded $G(s) = \mathrm{C(s)/R(s)}$ (fig. 3.28).

MATLAB also includes commands that allow the reduction of block diagrams automatically to get the single transfer function representative of the whole system.

Example 3.11

Reduce the block diagram of fig. 3.29 to a single transfer function by applying the necessary MATLAB commands for $G_1(s) = 10$, $G_2(s) = \frac{s+2}{s+10}$, $G_3(s) = \frac{1}{s+2}$, $H_1(s) = 5$, $H_2(s) = -2s$ and $H_3(s) = \frac{3}{s+2}$. Draw also the impulse response for the equivalent transfer function.

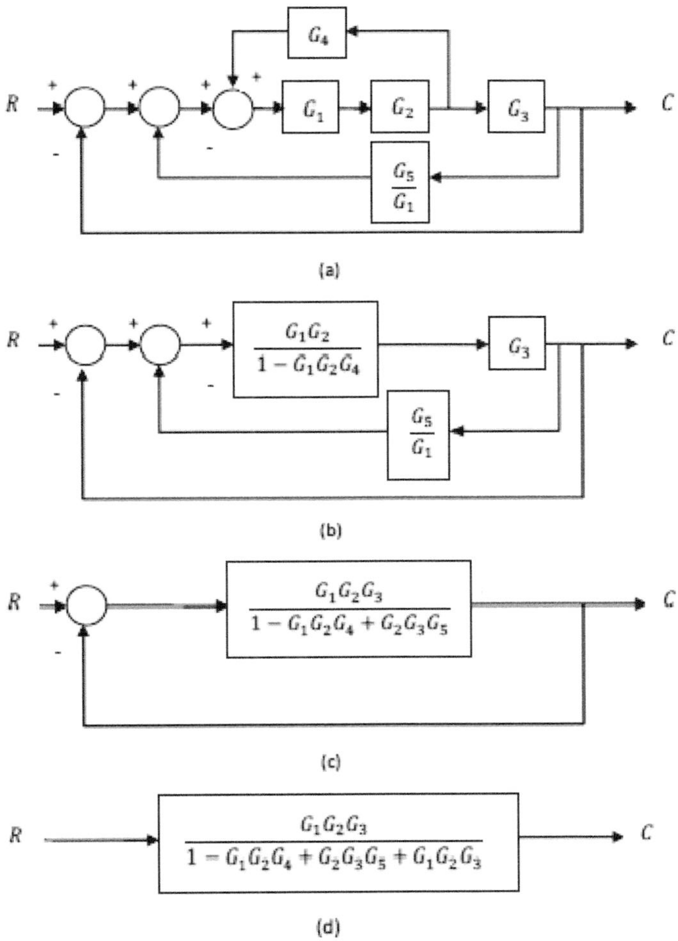

Fig. 3.28 Steps followed in reducing the block diagram for example 3.10 (from (a) to (d)).

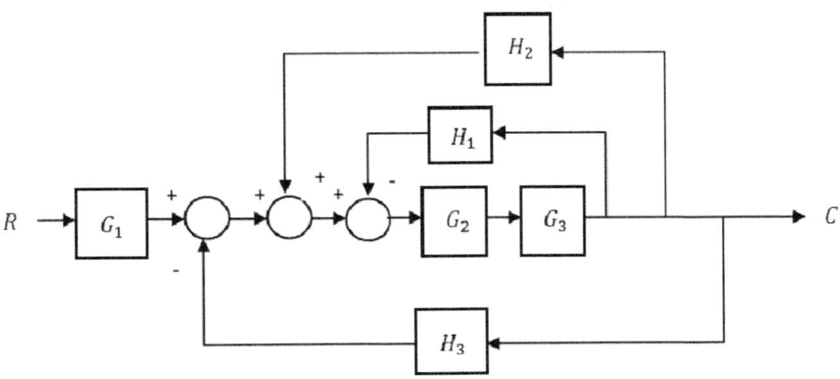

Fig. 3.29 Block diagram for example 3.11

Solution

In first place we define each one of the individual transfer functions

```
>> s=tf('s')
>> G1=10;
>> G2=(s+2)/(s+10);
>> G3=1/(s+2);
>> H1=5;
>> H2=-2*s;
>> H3=3/(s+2);
```

Then, we calculate the series product of $G_2(s)$ and $G_3(s)$

```
>> G23=G2*G3

Transfer function:
     s + 2
---------------
s^2 + 12 s + 20
```

We can group $H_1(s)$, $H_2(s)$, and $H_3(s)$ in a parallel way yielding

```
>> H123=H1-H2+H3

Transfer function:
2 s^2 + 9 s + 13
----------------
     s + 2
```

Below we reduce the negative feedback configuration formed by G23 and H123

```
>> Gfeed_G23_H123=feedback(G23,H123,-1)


Transfer function:
      s^2 + 4 s + 4
--------------------------
3 s^3 + 27 s^2 + 75 s + 66
```

Finally we compute the series product of the last transfer function with $G_1(s)$, yielding the reduced transfer function

```
>> G_reduced=G1*Gfeed_G23_H123

Transfer function:
    10 s^2 + 40 s + 40
--------------------------
3 s^3 + 27 s^2 + 75 s + 66
```

In fig. 3.30 the impulse response corresponding to the equivalent transfer function derived by using MATLAB is shown.

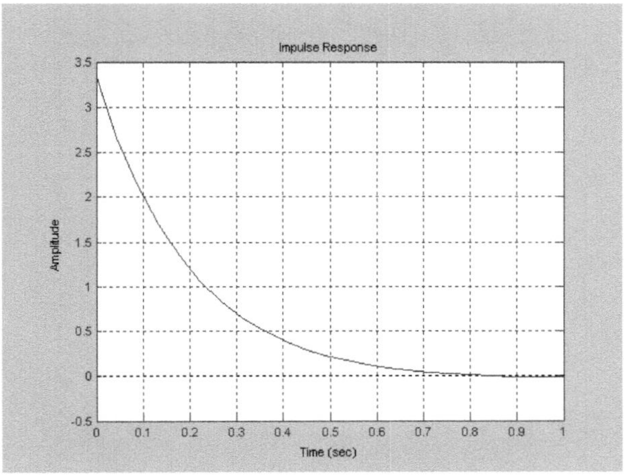

Fig. 3.30 Impulse response of the equivalent transfer function for example 3.11.

Chapter 4
System Response Analysis

This chapter analyzes the output of systems, also known as response, given a particular input. The utility of this study comes from the necessity of foreseeing how a physical system, e.g. the electromechanical steering system of a vehicle, reacts against an input stimulus, e.g. a sudden swerve. In this manner we can re-design the system at hand, probably adding new components, to make it to behave as desired.

The main characteristics of the response can be divided into two parts: the transient response and the steady-state response. The former entails the first reaction of the system, being the swiftness and smoothness of the response, the key features, while the latter refers to the permanent value of the output, being the error, i.e. the difference between the input and the response, the most important parameter.

In this chapter we will study analytically the response of first and second order systems and sketch the characteristic of the response of higher order systems based on their components.

4.1 System Time Response

The way in which physical systems react against input stimuli depends, obviously, on their own nature and the given input. When dealing with natural, non-modifiable systems, for instance objects in a spatial orbit, we can only observe and/or foresee the behavior of the system, for example the trajectory of a meteorite entering the atmosphere. Regarding human-made systems, however, it becomes essential to understand how systems react against different inputs in order to adapt them to fulfill the needed requirements. For example, first-order systems, like a thermometer, exhibit a characteristic curve response like the one shown in fig. 4.1 when they are exposed to a step input. According to the considered application we could require to modify the thermometer to achieve particular features. For instance, a medical thermometer should respond in few seconds with $\pm 0.1^\circ$ of precision. However, the temperature sensor of the heating system for an apartment can be designed to react more slowly, probably assuming a lower precision (which would make it cheaper).

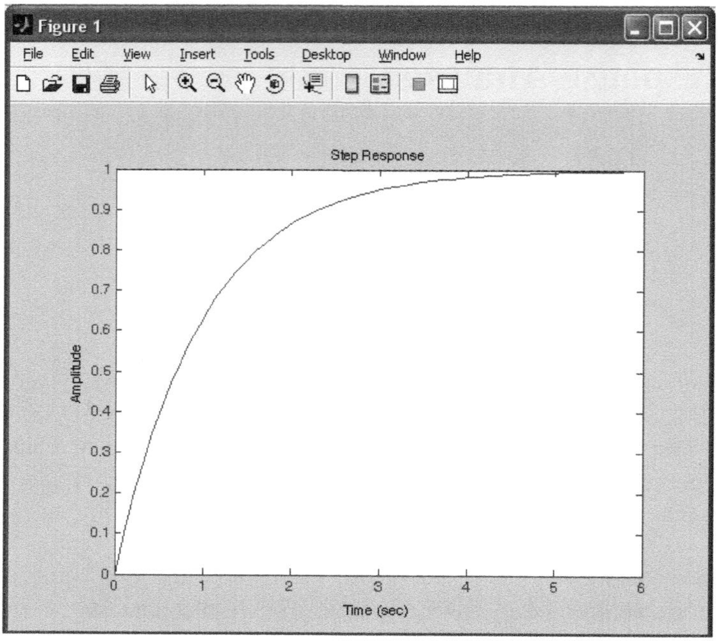


```
>>s=tf('s');
>>G=1/(s+1);
>>step(G);
```

Fig. 4.1 Response of a first-order system against step input. In this example the system follows the reference input pretty slowly, spending more than 4 seconds to achieve a final and stable value, which could be assumable for certain applications. Right, Matlab commands to obtain this response.

4.2 System Transient Response

Independent of the order of the system, we can distinguish between the natural response and the forced response. The natural response of a system depends on the system setup before the stimulus has been fired, that is, the response due to the initial conditions of the system, while the forced response regards to the system response due to the input stimulus considering the initial conditions as null. For example, if we study the speed of a vehicle when stepping on the accelerator, the forced response will account only for the speed increase, given the increment on the fuel volume, while the natural response will be the speed due to the initial velocity of the vehicle without considering the action on the accelerator.

Mathematically we can obtain the transfer function of both components of the system response. Given a general differential equation that models a system:

$$a_n y^{n)} + a_{n-1} y^{n-1)} + \ldots + a_1 y' + a_0 y = b_m u^{m)} + b_{m-1} u^{m-1)} + \ldots + b_1 u' + b_0 u \qquad (4.1)$$

The Laplace transformation considering initial conditions will be (recall chapter 3):

$$a_n\left(s^nY(s)-s^{n-1}y(0)-...-y^{n-1}(0)\right)+...+a_1\left(sY(s)-y(0)\right)+a_0Y(s)=$$
$$b_m\left(s^mU(s)-s^{m-1}u(0)-...-u^{m-1}(0)\right)+...+b_1\left(sU(s)-u(0)\right)+b_0U(s) \tag{4.2}$$

Grouping terms, we will obtain that the system response is the result of the sum of two components, lets say Y_1, the forced response and Y_2 the natural response which only contains the values of the initial conditions (noted here as P(s)).

$$Y(s)=\frac{b_ms^m+...+b_1s+b_0}{a_ns^n+...a_1s+a_0}U(s)+\frac{P(s)}{a_ns^n+...a_1s+a_0}=Y_1(s)+Y_2(s) \tag{4.3}$$

Example 4.1

Consider the following differential equation. Compute the natural and forced responses.

$$\ddot{y}+9\dot{y}+2y=3u \qquad y(0)=2,\ \dot{y}(0)=-4$$

Solution

Applying the Laplace transformation we obtain

$$s^2Y(s)-sy(0)-\dot{y}(0)+9(sY(s)-y(0))+2Y(s)=3U(s)$$
$$Y(s)(s^2+9s+2)=3U(s)+(9+s)y(0)+\dot{y}(0)$$
$$Y(s)=\frac{3}{s^2+9s+2}U(s)+\frac{2s+14}{s^2+9s+2}$$

fig. 4.2 shows a plot of the different parts of the response.

First-Order System Response

First-order systems come from differential equations like:

$$y'+a_0y=b_0u \tag{4.4}$$

For example, following the Fourier laws of thermal conduction, a normal thermometer can be modeled through the equation:

$$C\frac{dT}{dt}=\frac{T_r-T}{R} \tag{4.5}$$

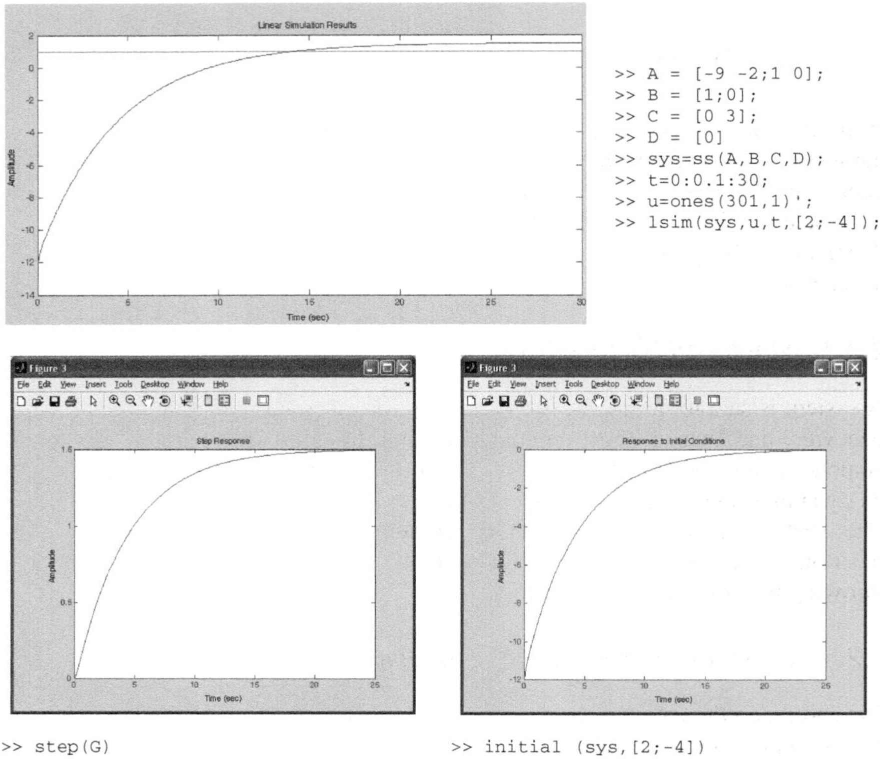

Fig. 4.2 System response against step input. Up) Response considering initial conditions. Down left) response without considering initial conditions (force response). Down right) Contribution of the initial conditions to the final response (natural response).

where C is the thermal capacitance of the thermometer, R the thermal resistance, t the measured temperature and t_r is the reference temperature. Thus, this equation can be rewritten to meet the expression (4.4) as:

$$\frac{dT}{dt}+\frac{1}{CR}T=\frac{1}{CR}T_r \tag{4.6}$$

with T, the system output and T_r the reference input.

In general, the transfer function of a first-order system obtained by means of the Laplace transformation of the equation (4.4) is:

$$G(s)=\frac{b_0}{s+a_0} \tag{4.7}$$

The response of the system, *Y(s)*, against step input with amplitude *A*, considering initial conditions as null, can be derived as follows:

$$Y(s) = \frac{b_0}{s+a_0}U(s) = \frac{b_0}{s+a_0}\frac{A}{s} \tag{4.8}$$

remembering that:

$$\ell^{-1}\{Au_e(t)\} = \frac{A}{s} \tag{4.9}$$

Applying residues we can separate this expression as:

$$Y(s) = \frac{K_1}{s} + \frac{K_2}{s+a_0} \tag{4.10}$$

with

$$K_1 = \frac{b_0 A}{a_0}; \quad K_2 = -\frac{b_0 A}{a_0} \tag{4.11}$$

Yielding that

$$Y(s) = \frac{b_0 A}{a_0}\left(\frac{1}{s} - \frac{1}{s+a_0}\right) \xrightarrow{\ell^{-1}} y(t) = \frac{b_0 A}{a_0}\left(1 - e^{-a_0 t}\right)u_e(t) \tag{4.12}$$

That is, the response of a first-order system against an abrupt change in the input (a step input) exhibits a negative exponential behavior as shown in fig. 4.1.

Notice that, assuming $a_0 > 0$, $\lim_{t\to\infty} e^{-a_0 t} = 0$, and thus, $\lim_{t\to\infty} y(t) = \frac{b_0 A}{a_0}$.

That is, the steady-state, i.e. the final value of a first-order response is $\frac{b_0 A}{a_0}$ and the term $\frac{b_0}{a_0}$ works as a factor that multiplies the amplitude of the input, so it is normally called *gain* and represented with *K*. Rewriting the system output using K, the expression for y(t) remains:

$$y(t) = KA\left(1 - e^{-a_0 t}\right)u_e(t) \tag{4.13}$$

Also note that the higher the magnitude of a_0, the faster the system response achieves the steady-state. We can define the *time constant* for first-order systems as $\tau = \frac{1}{a_0}$ which determines the moment in which the response is 63.21% of the final value. That is when t=τ,

$$\begin{gathered} y(t) = KA\left(1 - e^{-a_0 \tau}\right) \Rightarrow \{t = \tau\} \Rightarrow y(\tau) = KA\left(1 - e^{-a_0 \frac{1}{a_0}}\right) \Rightarrow \\ y(\tau) = KA\left(1 - e^{-1}\right) = \ 0.6321 \cdot KA \end{gathered} \tag{4.14}$$

The effect of the time constant τ on a first-order system can be observed in fig. 4.3. Now, in term of the parameters, K, and τ, the transfer function of a first-order system can be rewritten as:

$$\begin{gathered} G(s) = \frac{b_0}{s + a_0} \{\text{dividing by } a_0\}, G(s) = \frac{\frac{b_0}{a_0}}{\frac{1}{a_0}s + 1} \Rightarrow \\ G(s) = \frac{K}{\tau s + 1} \end{gathered} \tag{4.15}$$

that quickly identifies the system gain and the time constant.

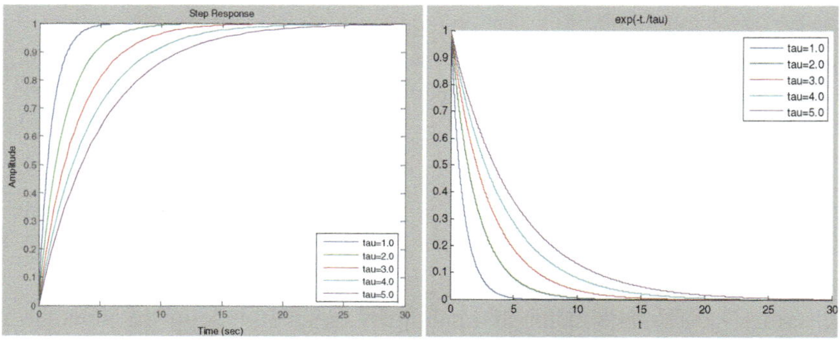

Fig. 4.3 System response for first-order systems. Left, step response varying the time constant. Lower values produce quicker responses. This is because the negative exponential term tends more quickly to zero (right).

Geometrically, the system gain and time constant are related to the tangent of the response at t=0. Deriving the time response (assuming a unitary step):

$$y(t) = K\left(1 - e^{-t/\tau}\right) \Rightarrow \frac{dy(t)}{dt} = \frac{K}{\tau} e^{-t/\tau} \Rightarrow \frac{dy}{dt}(0) = \frac{K}{\tau} \tag{4.16}$$

That is, we obtain that the derivate of the response at t=0 corresponds to, K/τ, which is the same

$$\tan(\alpha) = \frac{K}{\tau} \tag{4.17}$$

Fig. 4.4 illustrates this relation.

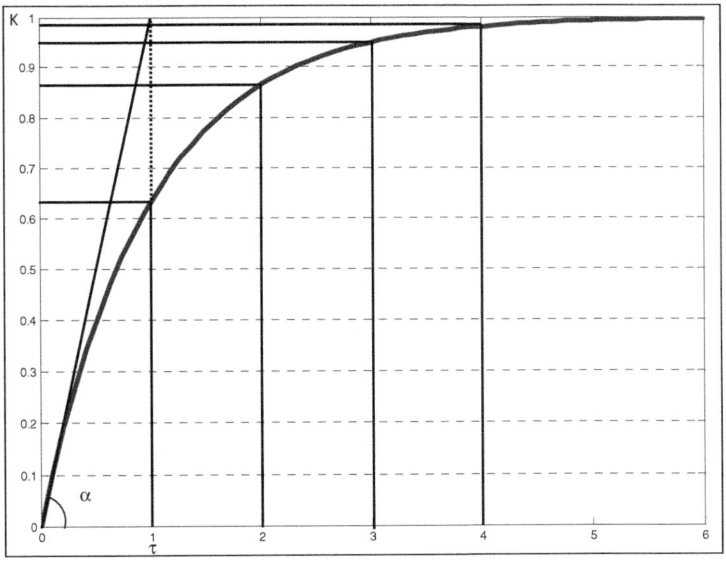

Fig. 4.4 System response for a first-order system with K=τ=1.

The reader may have noted that the coefficient a_0 has been considered as positive up to now. Having a negative value for this coefficient would make the system unstable since the exponential term of the response would grow indefinitely. That is, for $a_0<0$,

$$\lim_{t\to\infty} y(t) = \lim_{t\to\infty} \frac{b_0 A}{-a_0}\left(1 - e^{a_0 t}\right) = \infty \tag{4.18}$$

And the step output will be:

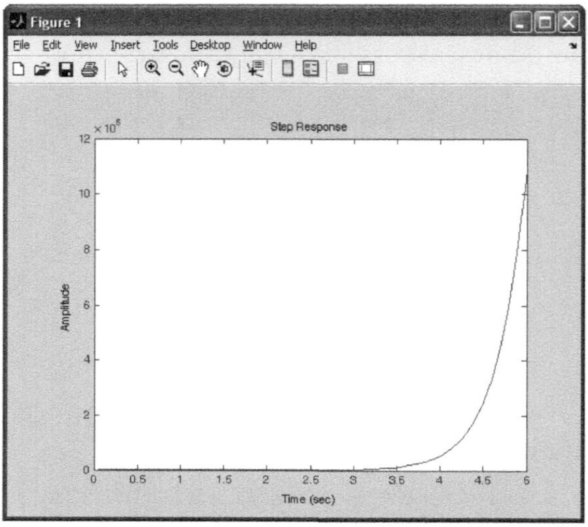


```
>>s=tf('s');
>>G=1/(s-3);
>>step(G);
```

Fig. 4.5 System response for an unstable first-order system.

Second-Order System Response

In general, second-order systems can be represented as differential equations in the form:

$$y''+a_1y'+a_0y=b_1u'+b_0u \tag{4.19}$$

That is, it relates the input (u(t)) and the output (y(t)) of the system involving, as maximum, the second derivation order of the latter. An example of a second-order system can be shown in fig. 4.6. This example is a simple representation of the shock-absorbing system of a vehicle made of a spring (with a force constant, *k*), a dashpot (with a damping constant, *b*), and the mass of the vehicle, *m*. Considering a rest setup at rest, (initial conditions considered as null) we are interested on studying the pose of the mass (y(t)) when a given input force (u(t)) is applied.

Recalling Chapter 2, and applying the force conservation principle, we can derive the differential equation shown in fig. 4.6, obtaining thus the transfer function for this system.

In general, the transfer function for a second-order system is[1]

$$G(s)=\frac{b_0}{s^2+a_1s+a_0} \tag{4.20}$$

where the characteristic equation (the denominator of the function) is of second order.

[1] Without generality loss, we assume there are no derivates for the input.

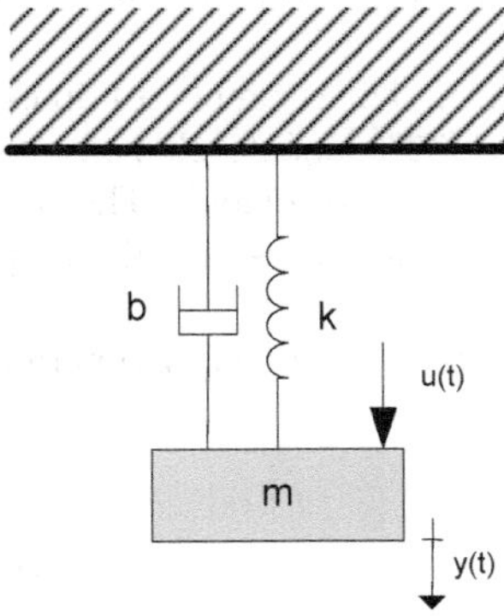

$$m\frac{d^2y}{dt}+b\frac{dy}{dt}+ky=f(t)\overset{\ell}{\Rightarrow}$$

$$ms^2Y(s)+bsY(s)+kY(s)=U(s)$$

$$Y(s)(ms^2+bs+k)=U(s)$$

$$G(s)=\frac{Y(s)}{U(s)}\Rightarrow G(s)=\frac{1}{ms^2+bs+k}$$

Fig. 4.6 A mechanical system and the derivation of its transfer function.

The response of a prototypal second-order system against a step input of amplitude A is

$$Y(s)=G(s)U(s)=\frac{b_0}{s^2+a_1s+a_0}U(s)=\frac{b_0}{s^2+a_1s+a_0}\frac{A}{s} \tag{4.21}$$

Applying residues we can separate this expression as:

$$Y(s)=\frac{K_1}{s}+\frac{K_2}{s+s_1}+\frac{K_3}{s+s_2} \tag{4.22}$$

And according to the nature of the three roots of the characteristic equation (also known as *poles*) we can distinguish three cases:

- s_1 and s_2 are both real and $s_1 \neq s_2$.
- s_1 and s_2 are real and $s_1 = s_2$.
- s_1 and s_2 are complex conjugates.

In the following we can discuss the particular responses for each case.

Second-Order System with Two Real, Different Poles

In this case, applying the inverse transformation of Laplace to equation (4.22) we obtain the general response expression for second-order systems when their poles are different and real.

$$y(t)=(K_1+K_2e^{-s_1t}+K_3e^{-s_2t})u_e(t) \tag{4.23}$$

Notice that this expression is similar to the response of first-order system but adding a new exponential term. The steady-state value for the response will be K_1 since the two exponential terms will tend to zero as long as the time tends to infinite as presented in the previous section.

Example 4.2
Consider the second order system:

$$G(s) = \frac{2.5}{(s+1)(s+2)}$$

Calculate the time response function when a unitary step is applied.

Solution

Given the transfer function, G(s), and the input U(s), the time response is calculated as:

$$Y(s) = G(s)U(s)$$

$$G(s) = \frac{2.5}{(s+1)(s+2)}, U(s) = \frac{1}{s}$$

$$Y(s) = \frac{2.5}{(s+1)(s+2)}\frac{1}{s} = \frac{K_1}{s} + \frac{K_2}{s+s_1} + \frac{K_3}{s+s_2}$$

When the input is a unitary step, one can derive the values for the residues K_1, K_2 and K_3 by using the Matlab command `residue`. Given the numerator and denominator of an expression, this command returns the ordered list of poles and their residues:

```
>> [r,p]=residue(2.5,conv([1 0],(conv([1 1],[1 2]))))
r =
    1.2500
   -2.5000
    1.2500
p =
    -2
    -1
     0
```

That is, in this case K_1=1.25, K_2=-2.5 and K_3=1.25, and therefore the output for this system is:

$$y(t) = (1.25 - 2.5e^{-t} + 1.25e^{-2t})u_e(t)$$

Graphically the response for this system is shown in fig. 4.7. Note that it is similar to the response of first-order systems, but a little bit slower at the beginning, appearing as a characteristic S-shape, which is normal due to the addition of an extra exponential term. Also, the magnitude of the poles is responsible for the time needed to achieve the steady-state: the smaller they are, the quicker the output reaches the final value. There is no need to say that s_1 and s_2 must be both positive to make the exponential terms to tend to zero. Otherwise the system will be unstable.

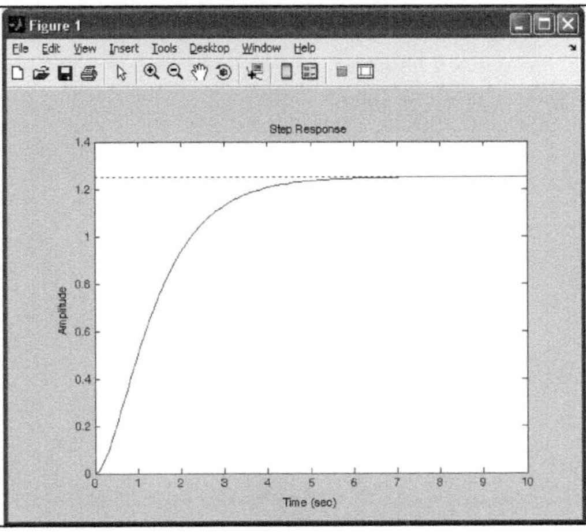


```
>> s=tf('s');
>> G=2.5/( (s+1)*(s+2))

Transfer function:
      2.5
-------------
s^2 + 3 s + 2

>> step(G)
```

Fig. 4.7 Second-order system response. The presence of two real (different) poles produces a smooth output. Recalling the final value theorem, it is easy to calculate the final value as 2.5/2=1.25.

Second-Order System with Two Real and Equal Poles

This case is very similar to the previous one with a response represented by (assuming unitary step input):

$$Y(s)=\frac{K_1}{s}+\frac{K_2}{s+s_1}+\frac{K_3}{(s+s_1)^2}\Rightarrow y(t)=(K_1+K_2e^{-s_1t}+K_3te^{-s_1t})u_e(t) \qquad (4.24)$$

The expression changes the exponential term by an exponential term multiplied by t, which will also tend to zero as long as time tends to infinite (assuming s_1 positive).

Example 4.3

Considering a modification of the previous example:

$$G(s)=\frac{2.5}{(s+1)(s+1)}$$

Calculate the time response function when a unitary step is applied.

Solution

Similarly to the Example 4.2

$$Y(s) = G(s)U(s)$$

$$Y(s) = G(s) = \frac{2.5}{(s+1)(s+1)}\frac{1}{s}$$

$$Y(s) = \frac{K_1}{s} + \frac{K_2}{s+s_1} + \frac{K_3}{(s+s_1)^2}$$

With K_2=K_3=-2.5 and K_1=2.5 given by:

```
>> [r,p]=residue(2.5,conv([1 0],(conv([1 1],[1 1]))))
r =
   -2.5000
   -2.5000
    2.5000
p =
    -1
    -1
     0
```

yields

$$y(t) = (2.5 - 2.5e^{-t} - 2.5te^{-t})u_e(t)$$

Regarding the response shape, it again exhibits similarities with respect to the case of two different poles, i.e. a S-shape at the beginning (see fig. 4.8). In this example, the final value is 2.5, i.e. the gain of the system is 2.5, as the reader can check by applying the final value theorem to

$$Y(s) = G(s)\frac{1}{s}$$

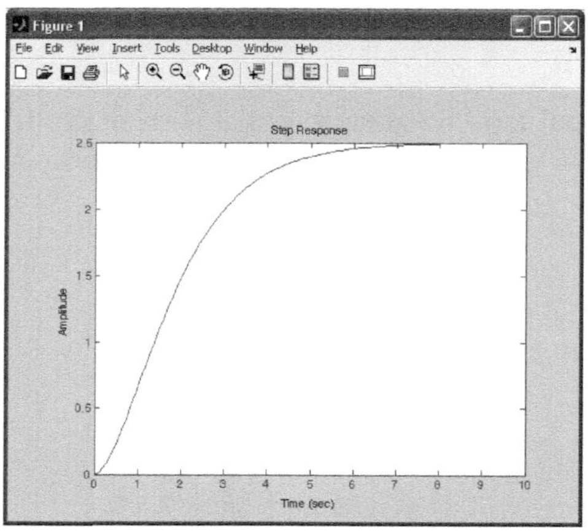


```
>> G=2.5/( (s+1)*(s+1))

Transfer function:
      2.5
-------------
s^2 + 2 s + 1

>> step(G)
```

Fig. 4.8 Second-order system response with two real and equal poles.

Second-Order System with Complex Conjugates

Up to now, the presence of real poles (different or equals) added an exponential term to the temporal response of the system. This notably changes when complex conjugate poles come into scene. The decomposition of the response of a second-order system with poles $s_{1,2} = \sigma \pm j\omega$ against step input yields:

$$Y(s) = \frac{K_1}{s} + \frac{K_2}{s+\sigma-j\omega} + \frac{K_3}{s+\sigma+j\omega} \tag{4.25}$$

which cannot be directly transformed to a temporal expression through the Laplace Transformation table. By means of a simple algebraic manipulation we can sum up the terms that involve the complex conjugates poles:

$$\begin{aligned} Y(s) &= \frac{K_1}{s} + \frac{K^*_2(s+\sigma)+K^*_3\omega}{(s+\sigma)^2+\omega^2} \\ K^*_2 &= 2\operatorname{Re}(K_2), \quad K^*_3 = -2\operatorname{Im}(K_3) \end{aligned} \tag{4.26}$$

And then, going back by separating the sum as:

$$Y(s) = \frac{K_1}{s} + \frac{K^*_2(s+\sigma)}{(s+\sigma)^2+\omega^2} + \frac{K^*_3\omega}{(s+\sigma)^2+\omega^2} \tag{4.27}$$

In this manner we obtain an equivalent expression in which each term can be directly transformed into:

$$y(t) = (K_1 + K^*_2 e^{-\sigma t}\cos\omega t + K^*_3 e^{-\sigma t} sen\omega t)u_e(t) \tag{4.28}$$

Analyzing this expression we note the presence of oscillations in the response due to the *sen* and *cos* terms. Also note that such oscillations will disappear as long as the time tends to be infinite, given that they are multipled by negative exponential terms.

As well as in previous cases the final value will be K_1 and in this case the effect of the oscillations will depend on the magnitude of the real part of the poles (σ) and their angular speed will be caused by the conjugate part (ω).

Example 4.4

Given the transfer function

$$G(s) = \frac{5}{s^2+4s+13}$$

Obtain the residues and the time response function.

Solution

$$Y(s) = G(s) = \frac{5}{s^2 + 4s + 13}\frac{1}{s}$$

$$Y(s) = \frac{K_1}{s} + \frac{K_2}{s+2+3j} + \frac{K_3}{s+2-3j}$$

The residues are:

```
>> [r,p]=residue(5,conv([1 0],[1 4 13]))

r =

  -0.1923 + 0.1282i

  -0.1923 - 0.1282i

   0.3846

p =

  -2.0000 + 3.0000i

  -2.0000 - 3.0000i

        0
```

That is, K_1=0.3846, K_2=-0.1923+0.1282j, and K_3=-0.1923-0.1282j. Therefore, K*2=-0.3846 and K*3=-0.257. With these values the response function is (see fig. 4.9 for a graphical representation):

$$y(t) = \left(0.3846 - 0.3846e^{-2t}\cos(3t) - 0.257e^{-2t}sen(3t)\right)u_e(t)$$

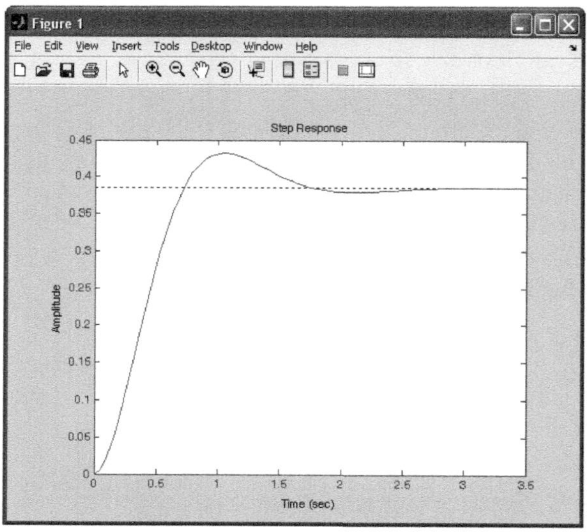


```
>> G=5/(s^2+4*s+13)

Transfer function:
       5
--------------
s^2 + 4 s + 13

>> step(G)
```

Fig. 4.9 Second-order system response with two complex conjugate poles. In contrast to previous cases, when the system has complex poles, the response presents a sinusoidal behavior.

Notice that in this case we have two parameters with influence in the response. On the one hand, as commented, the magnitude of the poles (the real part) is directly related to the time needed to achieve the steady-state, i.e. the final value. On the other hand, the complex part of poles (which can be considered as zero for the two first cases) is responsible for the frequency of the oscillations.

Fig. 4.10 illustrates the interrelation of both parameters studying the response of a second-order system against unitary input step.

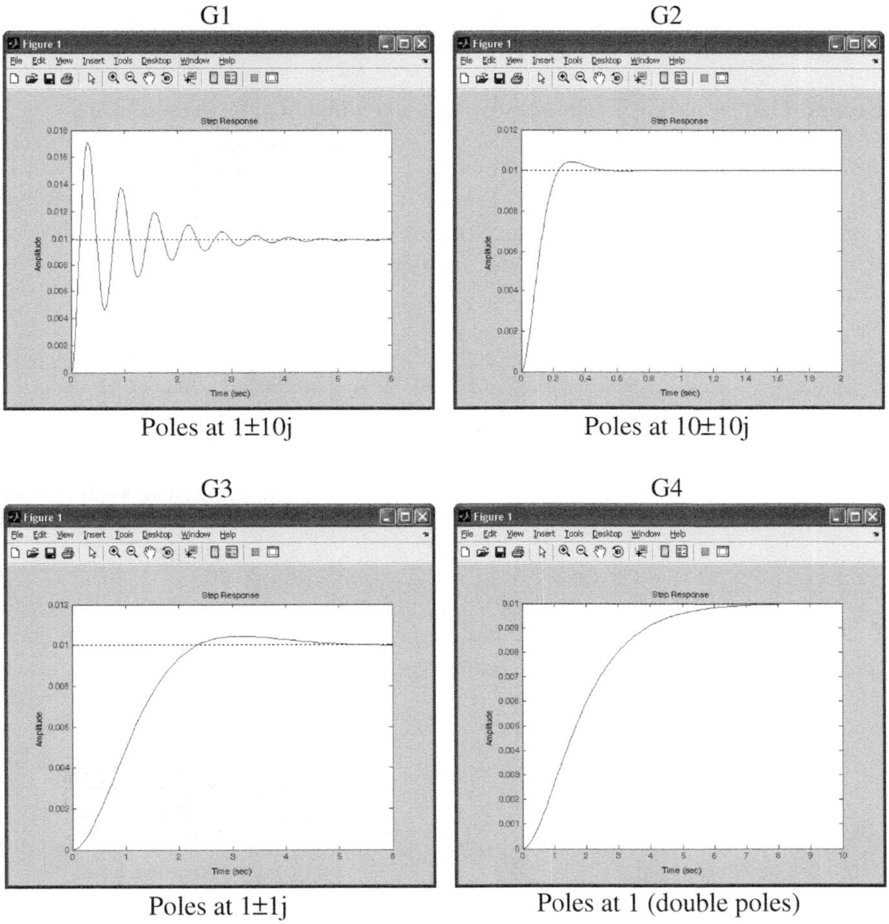

Fig. 4.10 Contribution of the real and complex parts of the poles in the response of second-order systems. Notice that the complex parts of poles affect the amplitude of the oscillations (see responses of G1, G3, and G4). Regarding the real part, the higher it is, the sooner the contribution disappear (compare responses of G1 and G2).

Comparing the outputs for G1 and G2, note that although they have poles with the same imaginary part, oscillations for G1 are pretty remarkable since the oscillations still remain after 5 seconds. G2, on the contrary, is quicker than G1 due to the difference of magnitude of the real part of the poles. This leads us to remember that the effect of poles far away from the ordinate axis of the s-plane is ephemeral. On the other hand, the responses of G1, G3, and G4 are good examples to realize the effect of the imaginary part of the poles in the amplitude of the oscillations. Systems with poles at the same real position but different imaginary components produces oscillations with different amplitude, including the absence of oscillation when no imaginary poles are considered (G4).

Second-Order System Response through a Geometrical Representation

Although the analysis and the representation used in the previous development is sufficient for understanding the behavior of second-order systems, a different representation oriented to geometrical variables is widely employed. The intuitive expression (4.29) is replaced with the more intriguing one given in (4.30).

$$G(s)=\frac{b_0}{s^2+a_1 s+a_0} \quad (4.29) \qquad G(s)=\frac{K\omega_n^2}{s^2+2\xi\omega_n s+\omega_n^2} \quad (4.30)$$

Since they are both equivalent, it is clear that

$$K\omega_n^2=b_0 \qquad 2\xi\omega_n=a_1 \qquad \omega_n^2=a_0 \quad (4.31)$$

In this representation, K is the gain of the system, ω_n is called the natural frequency and ξ is known as damping ratio. The roots of the characteristic equation (the poles of the system) can be obtained involving these variables as:

$$s_1, s_2=-\xi\omega_n \pm j\omega_n\sqrt{1-\xi^2}=-\sigma \pm j\omega_d \quad (4.32)$$

Where ω_d is called the damped frequency since it relates the damping ratio and the natural frequency.

Geometrically these variables are represented in the fig. 4.11. Trigonometrically it can be derived that the magnitude of the complex poles corresponds to ω_n and the phase, α, is $\cos^{-1}(\xi)$, i.e. $\xi=\cos(\alpha)$. Thus, given a second-order system expressed as in (4.30) one can guess the position of its poles

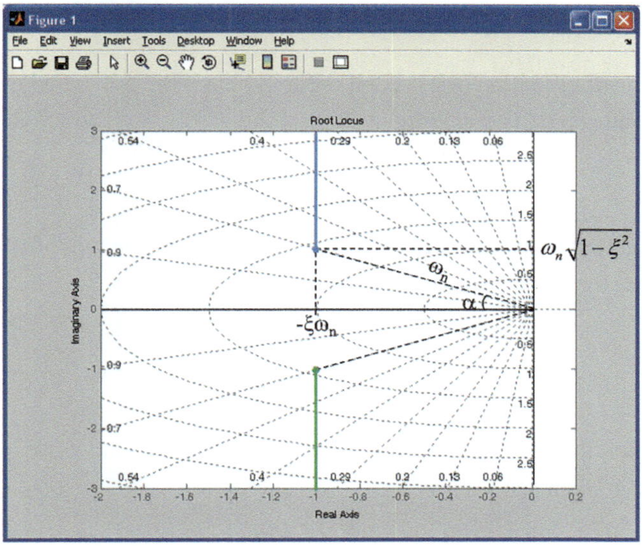

Fig. 4.11 Graphical meaning of the parameters ξ, ω_n, and ω_d

quickly according to the values of ω_n and ξ distinguishing the three cases previously commented:

1. $\xi > 1$

When $\xi > 1$, Expression (4.32) turns into two real numbers, that is, the poles of the system are real and different, and thus we are in the first case. Remembering that ξ is known as damping ratio, second-order system with $\xi > 1$ is called *overdamped.*

2. $\xi = 1$

When $\xi = 1$, the radical of Expression (4.32) is zero, and thus the poles are real and equal; we are in the second case. These systems are called *critically damped.*

3. $0 > \xi < 1$

In this case, poles are complex conjugates, and the system is called *overdamped.*

4. $\xi = 0$

This is an extreme case where poles are located at the imaginary axis, reaching the stability limit (remember that real part of poles has to be negative to assure the stability of the system). In this situation, the response of the system can be seen in fig. 4.12.

Example 4.5

Design a transfer function of a second order system with $\xi = 0$ and obtain its time response against a unitary step input.

Solution

Assuming $\xi = 0$ in the the general form of the transfer function of a second order system given by the equation (4.30), we obtain

$$G(s) = \frac{K\omega_n^2}{s^2 + \omega_n^2}$$

For example, for K=1 and $\omega_n^2 = 10$, the following Matlab code produces the system output, depicted in fig. 4.12.

```
>> s=tf('s');
>> G=10/(s^2+10)

Transfer function:
   10
--------
s^2 + 10

>> step(G)
```

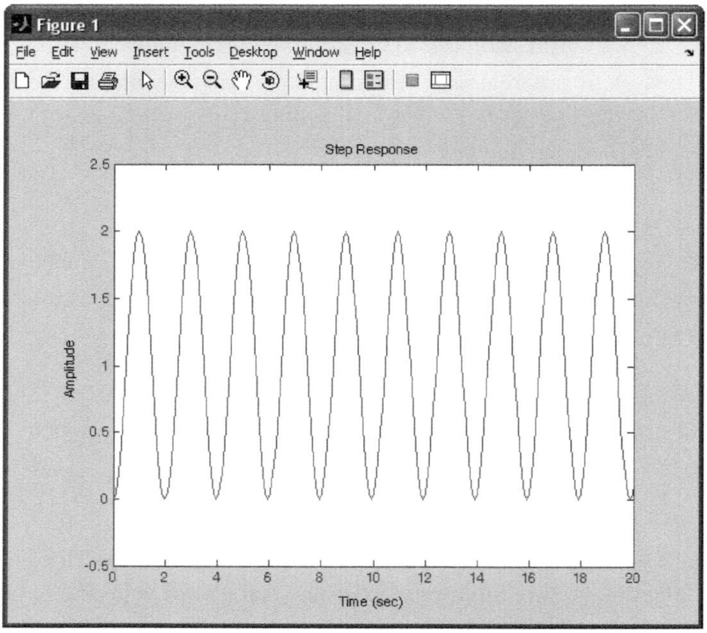

Fig. 4.12 Second-order system response with $\xi = 0$.

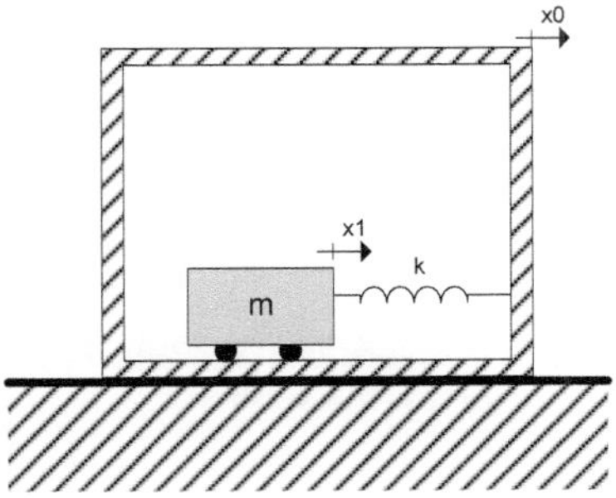

a)

$$m\ddot{x}_1 + k(x_1 - x_0) = 0$$

$$\{y = x_1 - x_0 \Rightarrow x_1 = y + x_0 \Rightarrow \ddot{x}_1 = \ddot{y} + \ddot{x}_0\}$$

$$m\ddot{y} + m\ddot{x}_0 + ky = 0$$

$$m\frac{d^2y}{dt^2} + ky = -m\frac{d^2x_0}{dt^2}$$

$$G(s) = \frac{Y(s)}{X_0(s)} \Rightarrow$$

$$ms^2Y(s) + kY(s) = -ms^2X_0(s)$$

$$G(s) = \frac{-ms^2}{ms^2 + k}$$

b)

```
G=-10*s^2/(10*s^2+5)

Transfer function:
 -10 s^2
----------
10 s^2 + 5

>> step(G)
```

c)

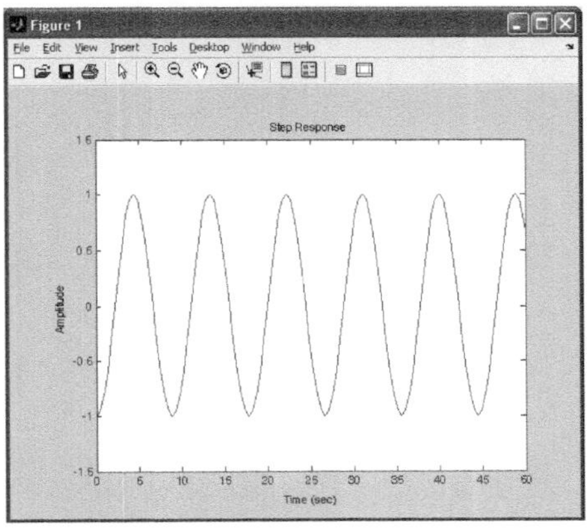

d)

Fig. 4.13 Cart position after a unitary step input. Given that no friction has been modeled, the cart will never stop.

Example 4.6

Fig. 4.13 a) depicts a cart inside a box and connected to it through a spring. The box is placed on the ground in such a way that any movement of the ground will be transmitted to the box and the cart will start to move. It can be considered as a naïve seismographer. Obtain the transfer function for this system and its step sponse.

Solution

The equations and the transfer function that model this system are given in fig. 4.13 b) while fig. 4.13 c-d) depicts the step response of the system and the Matlab code. The reader will be wondering why the cart is eternally oscillating when the input has been a unitary step. The answer is simple; in this example cart friction has not been considered and thus is moving indefinitely since no forces counteract the movement.

To sum up, the fig. 4.14 entails the exposed casuistry for second-order systems

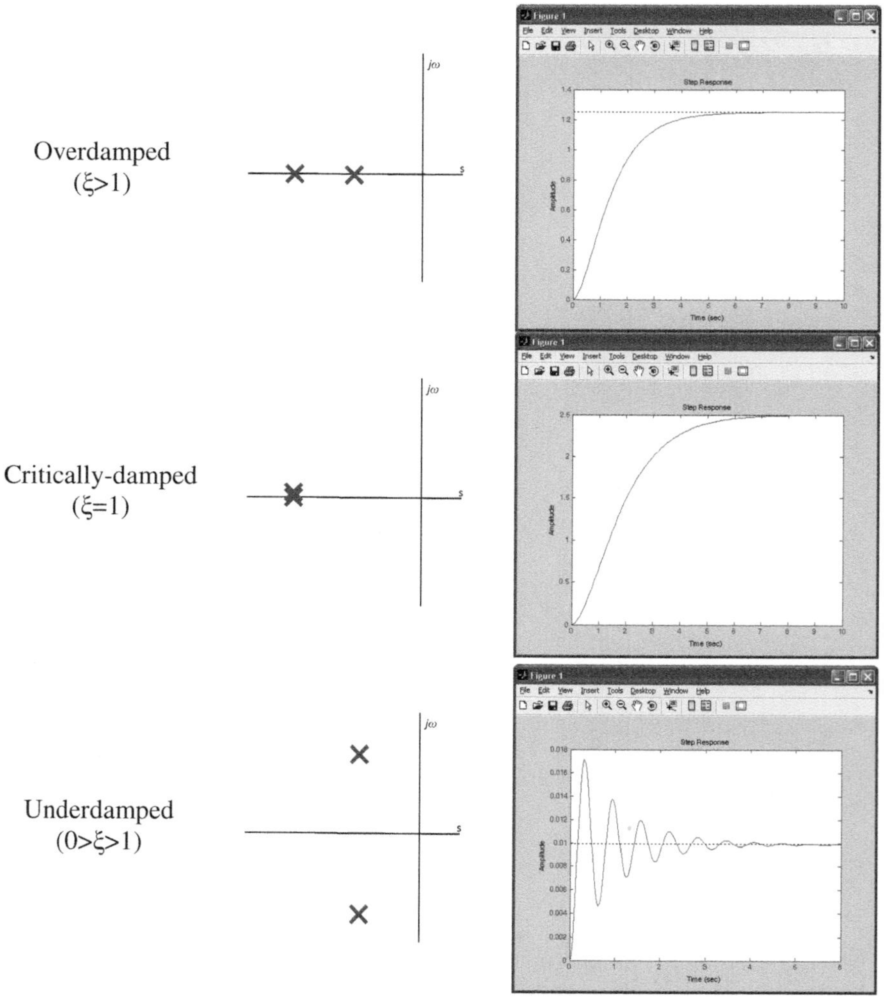

Fig. 4.14 Typical responses for second-order systems according to the value of ξ.

Underdamped
ξ=0

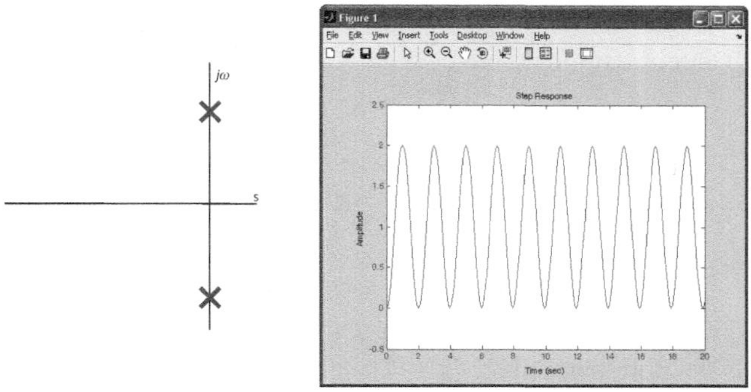

Fig. 4.14 (*continued*)

Higher-Order System Response

Unfortunately (or not), real systems are not so trivial than first- or second-order systems and the study of their responses cannot be described in detail as we did in the previous section. The casuistry will depend obviously on the order of the system, i.e. the number of poles, and the number of zeroes. However, a generalized study can be derived. In general, an n-order-system can be represented as:

$$G(s)=\frac{K(s+z_1)(s+z_2)...(s+z_m)}{(s+p_1)(s+p_2)...(s+p_n)} \tag{4.33}$$

Without thoroughness loss, we can assume that the n poles of the transfer function (4.33) are real and different and that the input signal is a unitary step. With these assumptions the response Y(s) can be separated as:

$$Y(s)=G(s)\frac{1}{s}=\frac{K_0}{s}+\sum_{i=1}^{n}\frac{K_i}{s+p_i} \tag{4.34}$$

which yields the temporal expression

$$y(t)=(K_0+\sum_{i=1}^{n}K_ie^{-p_it})u_e(t) \tag{4.35}$$

Note that the response of a second-order system with real and different poles is a particular case of Expression (4.35), that is, assuming $p_i>0$, the system output will tend to the final value K_0. The transient response will depend on the magnitude of the poles and the values of the coefficients K_i.

When analyzing high-order systems the concept of *dominance of poles* appears. It can be said that pole p_1 dominates pole p_2 if the following rule yields[2].

$$\frac{\mathrm{Re}(p_2)}{\mathrm{Re}(p_1)} > 5 \tag{4.36}$$

Example 4.7

Given a second-order system with poles at $p_1=1$ and $p_2=10$. Determine which pole is the dominant one.

Solution

The pole $p_1=1$ dominates $p_2=10$ and therefore the effect of the latter on the response can be insignificant.

Fig. 4.15 illustrates this example. It depicts the response of a first-order system with $p_1=1$ (in red) and the response of a second-order system with $p_1=1$ and $p_2=10$. Notice that the effect of p_2 is insignificant and the second-order system can be considered as a first-order one, neglecting p_2.

Through the dominance of poles we can, when applicable, reduce the order of systems and thus facilitating its study.

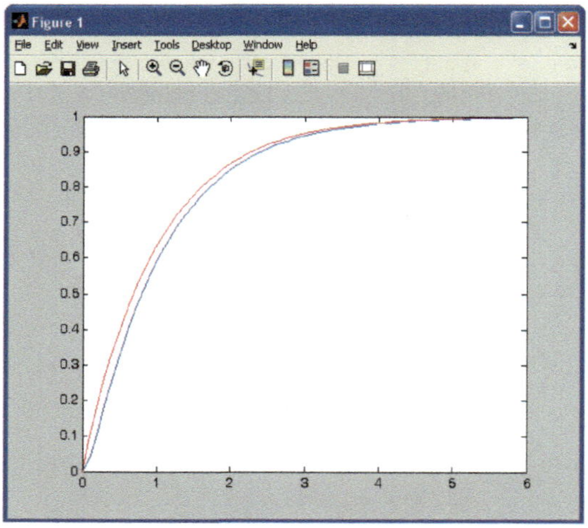


```
>> s=tf('s');
>> G=10/( (s+1)*(s+10))
>> G2=1/(s+1)
>> [x,y]=step(G);
>> [x2,y2]=step(G2);
>> plot(y,x)
>> hold on
>> plot(y2,x2,'r')
```

Fig. 4.15 Dominance of poles. The response of the second-order system with poles at s=-1 and s=-10 (blue) is quite similar to the one of a first-order system with a pole at s=-1 (red). This is due to the slight contribution of the pole s=-10 in comparison with the one at s=-1, which is called a dominant pole in this case.

[2] In absence of zeroes near p_1 or p_2.

In the general case of dealing with conjugate complex poles, similar considerations are applicable.

Fig. 4.16 shows the contribution to poles in the response to a high-order system according to their relative position. The final output will be the sum of their contribution along the time.

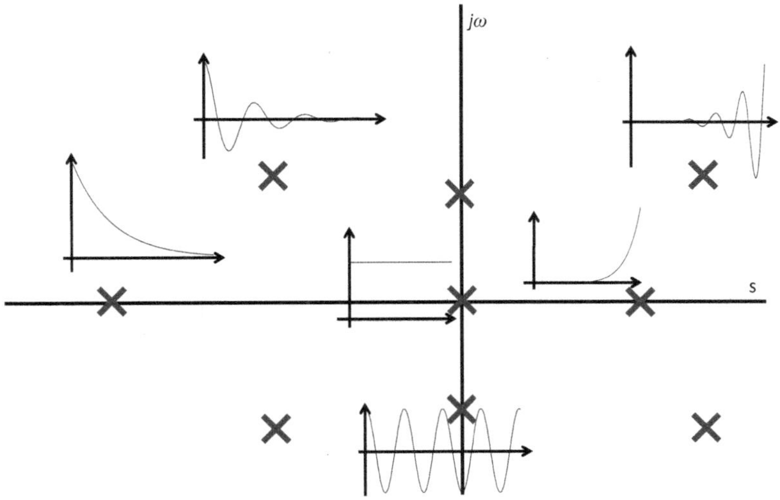

Fig. 4.16 Contribution of poles according to their relative position at the s-plane. Notice that poles located at the right part of the s-plane produce unstable contributions (tending to infinite).

4.3 Root Locus

Previous sections have highlighted the particular contribution of poles of a system according to their position. Thus, when analyzing the response of a given system we can infer its order and the position of its poles. In a certain way, we are now interested on the opposite: we want to change the response of a given system, i.e. its behavior, by *moving* its poles as desired. To do that we rely on a close-loop configuration including a parameter whose value will alter the original system.

The root locus technique was proposed by Evans in 1948 as a graphical representation of the position of the poles of a close-loop system according to the variations of a system parameter. The general configuration is shown in fig. 4.17, where the output of the system G(s) is compared to the input, through the H(s) system. The error E(s), i.e. the difference between the input and the meassured output, is modulated by the parameter K, being E(s)*K the input of G(s).

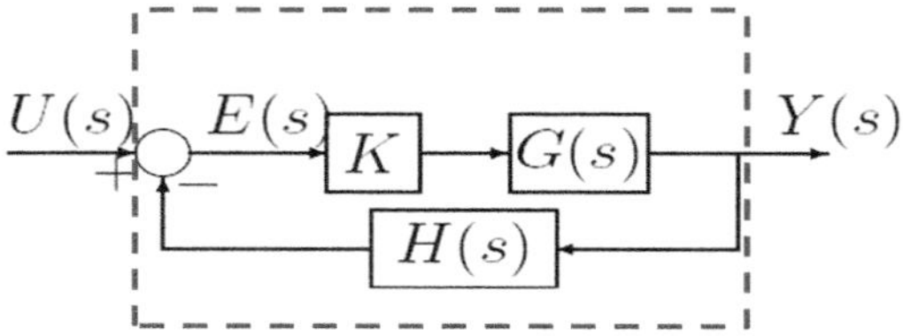

Fig. 4.17 Close-loop configuration. Through the root locus, all the possible locations of the poles of the system can be calculated in function of the parameter K.

This is actually a typical *proportional control* where G(s) is the plant to be controlled, H(s) represents a sensorial system that measures the plan output, and K is a gain that amplifies the error signal given to G(s). When error is zero, in case it happens, it means that Y(s)=U(s), and the input to the system G(s) is also zero, therefore, no actuation is carried out.

Chapter 5 goes in detail on proportional and other control systems, by now we are going to cope with the behavior of the close-loop system according to the parameter K. This behavior, as commented before, will be defined by the position of poles, and the root locus technique comes to depict their possible locations.

Recalling Chapter 3, the equivalent transfer function of the close-loop configuration of fig. 4.17 is:

$$G_{tot}(s) = \frac{KG}{1+KGH} \tag{4.37}$$

being the characteristic equation

$$1+KG(s)H(s)=0$$

In this equation the parameter K, and thus, variations in its value will modify its roots, and therefore the position of the poles.

Example 4.8

Considering

$$G(s) = \frac{1}{s^2+7s+10} \qquad H(s)=1$$

Obtain the close-loop transfer function and the position of its poles according to different values of K.

Solution

The close-loop transfer function is:

$$G(s) = \frac{K}{s^2 + 7s + 10 + K}$$

Giving values to the gain K and calculating the roots of the characteristic equation for this second-order system, we obtain

K	P1	P2
0	-5	-2
1	-4.62	-2.38
2	-4	-3
2.25	-3.5	-3.5
3	-3.5+0.872i	-3.5-0.872i
6.5	-3.5+2.06i	-3.5-2.06i

From this we can deduce that if $0 \le K \le 2.25$ the system is overdamped, that is it has different and real poles, for K=2.25 the system is critically damped, i.e. it has two real and equal poles, finally if $K \ge 2.25$, the system is underdamped, exhibiting a oscillating behavior due to the presence on complex conjugated poles. If we depict the position of the poles as long as the value of K is increasing we obtain:

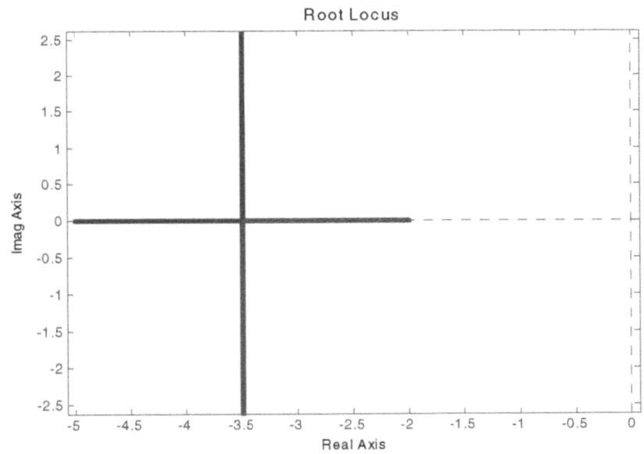

Fig. 4.18 Root Locus example. Illustration of how the poles *move* as long as the parameter K increases.

That is, we can rapidly observe the position of each pole, and therefore inference the behavior of the final system, in a plot like the one shown in fig. 4.18. Obviously relying on a systematic calculation of the roots of the characteristic equation to depict the possible position of the poles based on a given parameter is not practical at all and a set of recipes are available for sketching it.

4.3.1 Root Locus Recipe

There exists a set of simple steps for sketching the root locus of a given system. They are mathematically derived from the complex nature of the characteristic equation. That is, given expression (4.38), we can distinguish the real and complex part of the equation and rewrite it as shown in (4.39).

$$1+KG(s)H(s)=0 \Rightarrow KG(s)H(s)=-1 \tag{4.38}$$

$$|KG(s)H(s)|\angle KG(s)H(s)=-1+j0 \tag{4.39}$$

That is,

$$|KG(s)H(s)|=1 \tag{4.40}$$

$$\angle KG(s)H(s)=\angle -1=\pi(2k+1), \quad k=0,\pm1,\pm2,... \tag{4.41}$$

Equation (4.40) is known as the magnitude condition and can be used to calculate the value of the parameter K for a particular position of the poles. Equation (4.41) is called the angle condition and permits us to determine whether a particular point of the s-plane belongs or not to the root locus, i.e. if there exist a value for K for which one of the poles of the system are located at that particular point.

These two equations are the basis for the root locus method.

Example 4.9

Check the application of the magnitude and angle conditions to the system

$$G(s)=\frac{1}{s^2+7s+10} \qquad H(s)=1$$

At the test point s, $s_1=-3.5+j$.

Solution

G(s)H(s) can be rewritten as:

$$G(s)H(s)=\frac{1}{(s+2)(s+5)}$$

For the test point (see fig. 4.19), angle condition is computed as:

$$\angle\frac{K}{(s+2)(s+5)}\bigg|_{s=s_1}=-\angle(s_1+2)-\angle(s_1+5)=-[(\pi-\theta)+\theta]=-\pi$$

And therefore this condition holds for this test point. Actually it holds for

$$s_1=-3.5\pm xj, \quad x\in[0,\infty)$$

Now we have confirmed that s_1=-3.5+j belongs to the root locus we want to figure out the value of K that moves one of the poles to this position. For that we rely on the magnitude condition:

$$\left|\frac{K}{(s+2)(s+5)}\right|\Bigg|_{s=s_1} = \frac{K}{|s_1+2||s_1+5|} = 1 \Rightarrow K = |s_1+2||s_1+5| = 3.6$$

Yielding the value of K=3.6, which confirms what we figure out by sequentially solving the characteristic function for different values of K.

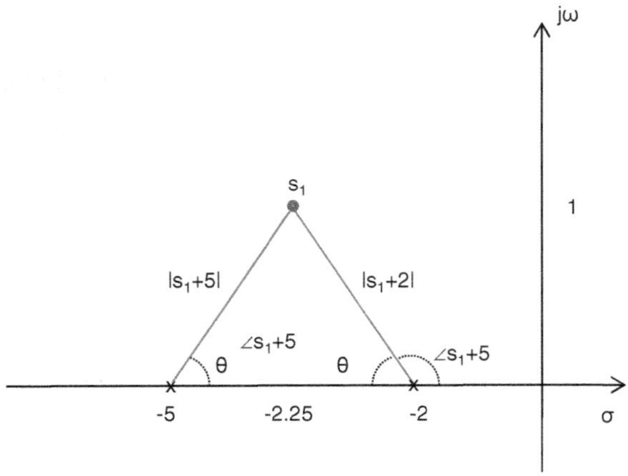

Fig. 4.19 Root Locus example where the point s=s_1 is being checked whether it belongs or not to the root locus (more description in the text).

In general, for an n-order system, the characteristic equation is given by $F(s) = -1$ being

$$F(s) = \frac{K(s+z_1)(s+z_2)\ldots(s+z_m)}{(s+p_1)(s+p_2)\ldots(s+p_n)} \tag{4.42}$$

and the magnitude and angle condition computed as:

$$\angle F(s) = \sum_1^m \angle(s+z_i) - \sum_1^n \angle(s+p_i) = \pi(2k+1), \quad k = 0,1,2,\ldots \tag{4.43}$$

$$|F(s)| = \frac{\prod_1^m |s+z_i|}{\prod_1^n |s+p_i|} = 1 \tag{4.44}$$

Next, a list of steps are given based on these conditions for sketching the root locus.

Step 1. Placement of the poles and zeroes of G(s)H(s) on the s-plane

Firstly it is recommendable to rewrite the equation

$$1+K\frac{\prod_{1}^{m}(s+z_i)}{\prod_{1}^{n}(s+p_i)}=0\Rightarrow\prod_{1}^{n}(s+p_i)+K\prod_{1}^{m}(s+z_i)=0 \qquad (4.45)$$

Mark on the s-plane the poles p_i with a cross 'x' and the zeroes z_i with a small circle 'o'. Note that when K=0, equation (4.45) is reduced to

$$\prod_{1}^{n}(s+p_i)=0 \qquad (4.46)$$

And thus, the poles of the closed-loop system are identical to the poles in the open-loop configuration, that is G(s)H(s). Also notice that as long as K tends to ∞, the roots of the equation (4.45) coincides with the zeroes of G(s)H(s):

$$\begin{aligned}&\frac{1}{K}\prod_{1}^{n}(s+p_i)+\prod_{1}^{m}(s+z_i)=0\\&\lim_{K\to\infty}\left(\frac{1}{K}\prod_{1}^{n}(s+p_i)+\prod_{1}^{m}(s+z_i)\right)=\prod_{1}^{m}(s+z_i)\end{aligned} \qquad (4.47)$$

Thus, at this first step we obtain significant information about the root locus sketch: Root locus starts at the poles of the open-loop system, i.e. the root locus will be made of n strokes, each one starting at a pole of G(s)H(s), when K=0, and some of them, precisely m, ending at the zeroes of G(s)H(s), when K→∞.

Step 2. Root locus on the real axis

In this step we focus on which part of the real axis will be part of the root locus, that is we are going to apply the angle condition to points $s\in\Re$, and s=(-∞,∞). A practical rule to remember tells us that "a given point s^* lying on the real axis belongs to the root locus if and only if the sum of the real poles and real zeroes place on its right is an odd number". This fanciful rule is based on the angle condition and the need of yielding an odd multiple of π.

Example 4.10

Sketch the root locus over the real axis for the open-loop system

$$G(s)H(s) = \frac{1}{s(s+1)(s+2)}$$

Solution

After step 1, we locate the three poles at s=0, s=-1, and s=-2 as shown in fig. 4.20, and test a point s between p_1 and p_2, that is s∈(-2,-1). In absence of zeroes, the angle condition is computed, in this example, as

$$0-\sum_{1}^{3}\angle(s+p_i)=\pi+\pi+0=2\pi\neq\pi(2k+1),\quad k=0,1,2,...$$

Note that the presence of two poles on the right of s contributes with 2*π, while the pole on the left does not contribute. Thus, it is obvious that s is not a part of the root locus, as well as any point in the interval (-2,-1). Following the same argument, ∀s∈(-∞,-2] and ∀s∈[1,0], *s* is part of the root locus.

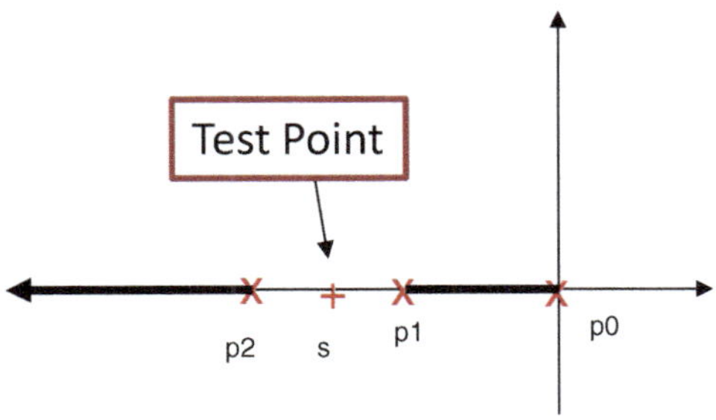

Fig. 4.20 Testing points on the real axis. Based on the angle condition, a point on the real axis will belong to the root locus if the number of poles (and zeros) placed on its right is odd. In this case, the *s* point does not belong to the root locus.

In this example, we observe that a certain direction is given to one segment of the root locus to indicate the path followed by the pole p2 as long as the parameter K increases. Thus, p2 tends to -∞ and it can be said that the root locus entails a segment that starts at p2 and ends at -∞.

Step 3. Assessing asymptotes

In step 1 we stated that the segments of the root locus begins at the poles of G(s)H(s) when K=0 and ends at the zeroes when K=∞. Normally, in causal systems, the number of zeroes, *m*, is less than the number of poles, *n*, and thus only *m* segments will end at a zero. The rest of segments, concretely *m-n* will tend asymptotically to infinity. In the example of fig. 4.20, *m*=0 and *n*=3, i.e. there are no zeroes, and thus the three segments that start at each pole asymptotically go forward to infinity. The angle of the asymptotes with respect the real axis can be determined by the equation (4.48).

$$\phi = \frac{\pm\pi(2k+1)}{n-m} \quad k = 0,1,2... \tag{4.48}$$

This expression comes from the application of the angle condition to a distant point (ideally at the infinity). At that distant point the angles from poles and zeroes can be assumed as equal and thus, they are cancelled, turning the angle condition into

$$\phi = \frac{\pi}{n-m}$$

The place where the asymptotes cross the real axis is determined by the mass center of the poles and zeroes as:

$$s = \sigma_a = \frac{\sum_{i=1}^{n}(-p_i) - \sum_{i=1}^{m}(-z_i)}{n-m} \tag{4.49}$$

In the example of fig. 4.20 there are two asymptotes, with $\sigma = \pm\pi/2$, located at $\sigma_a=(-2-5)/2=-3.5$.

Step 4. Identification of rupture points

In some occasions, two or more segments of the root locus collide at a particular point of the s-plane, which is called a *rupture point*. That is, a rupture point occurs when, given a certain value of K, the characteristic equation has roots with a multiplicity greater than 1. Identifying these points will largely help to sketch the root locus further away from the real axis.

Rupture points coincide with the maximum values of K so in practice, they are computed by deriving K and solving the equation dK/ds=0. The solutions of this equation, s*, fulfilling that K(s*)>0 are rupture points.

Example 4.11

Calculate the rupture points for the system

$$G(s) = \frac{1}{s^2+7s+10} \qquad H(s) = 1$$

Solution

The characteristic function is:

$$s^2 + 7s + 10 + K = 0$$

from which

$$K = -s^2 - 7s - 10 \Rightarrow \frac{dK}{ds} = -2s - 7$$

And solving

$$\frac{dK}{ds} = 0 \Rightarrow -2s - 7 = 0 \Rightarrow s = -3.5$$

Checking that it is really a maximum of the curve K, we have that

$$K = -\left(-3.5^2\right) - 7(-3.5) - 10 > 0$$

and thus s=-3.5 is a rupture point as revealed in the fig. 4.18.

Although it is common to find rupture points lying on the real axis, there are some cases where the roots of dK/ds are complex conjugates as illustrated in fig. 4.21.

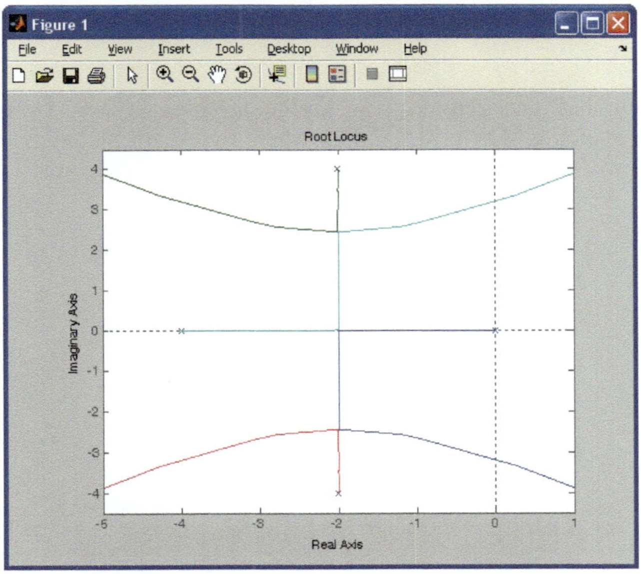

Fig. 4.21 Root Locus example for a system with 4 poles. It is made of four segments that collides at three rupture points: s=-2 and s=-2±2.45j

Step 5. Determination of starting/ending angles

This step is a straightforward application of the angle condition. It provides us with valuable information about the direction of the segments when starting at the poles or when ending at the zeroes.

$$\varphi_{pole} = \pi - \sum_{rest} \varphi_{poles} + \sum_{rest} \varphi_{zeroes}$$
$$\varphi_{zero} = \pi - \sum_{rest} \varphi_{zeroes} + \sum_{rest} \varphi_{poles} \tag{4.50}$$

Step 6. Crossing points of the root locus with the complex axis

This is the last step for sketching a root locus and it is intimately related to stability. In fact, in this step we calculate the value of the parameter K that made unstable the close-loop system. At the moment, at least one pole of the close-loop system will be located at the complex axis.

This step is not always applicable since there are cases where the system is always stable regardless of the value of K[3], as it is the example of fig. 4.18.

Example 4.12

Sketch the root locus for the system

$$G(s)H(s) = \frac{s+1}{s(s+4)(s^2+2s+2)}$$

Solution

This system has one zero at s=-1 and four poles at s=0, s=-4 and s=-1±j. The initial sketch is shown in fig. 4.22 where two segments lay on the real axis, from s=-4 to infinity and from s=0 to s=-1and three asymptotes (*n*-*m*=3) with

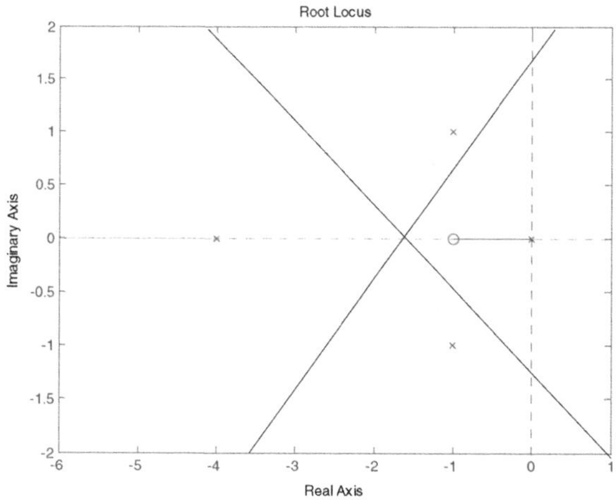

Fig. 4.22 Root Locus example where two segments (those over the real axis) has been traced. The asymptotes indicate the direction followed by the other two segments, starting at the complex poles.

[3] When K defined in the range [0,infinity).

angles 60,180 and 300°, crossing the real axis at σ_a=-5/3. At this moment we realize that there should be two segments more, starting at s=-1±j and that they have to tend to infinity following the asymptotes, but how? Calculating the starting angle from these poles we obtain ±26.6°, giving us a cue about their orientation.

The sketch of the root locus is finished when the stability of the system and the value of K and the location of the poles at the limit of the stability is computed. In this example the Routh criteria yields that the system is unstable when K>24.78. For this precise value, the characteristic equation has two roots at s=0±2.33j. Therefore the root locus crosses the complex axis at ±2.33j. fig. 4.23 depicts the root locus computed by the Matlab command `rlocus` that confirms our initial sketch was correct.

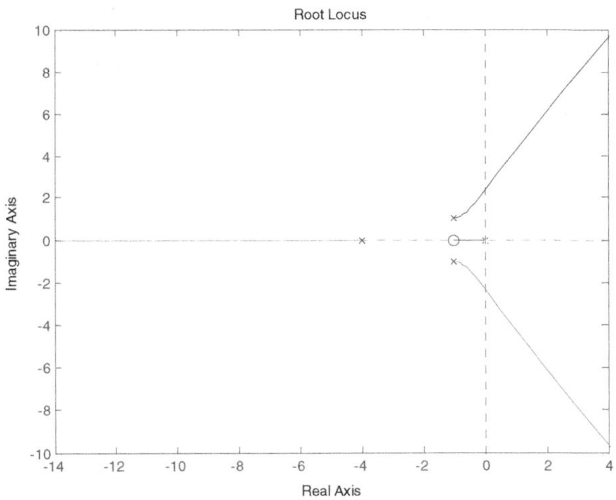

Fig. 4.23 Root Locus example depicted by rlocus. Note that there is a value of K that brings the poles of the system to the right part of the s-plane, and thus, makes the system unstable.

In the following section the effect of adding/removing poles zeroes to the open-loop system is commented.

4.3.2 *Effects of Adding Poles/Zeros*

Sometimes we are interested in modifying the behavior of a close-loop system to obtain a particular response, e.g. reducing the overshoot, or to gain in stability. One simple manner to do this is to add/remove poles/zeroes from the initial configuration up to achieving our goal. In this section we analyze the effect of this operation in the root locus.

Let's begin with the simplest root locus, the root locus of a first-order system. fig. 4.24 shows the root locus for a system with a pole at s=-5 and the close-loop response for different values of K. Notice that as long as K increases, the pole moves over the negative part of the real axis at the same time that the output reduces the permanent error.

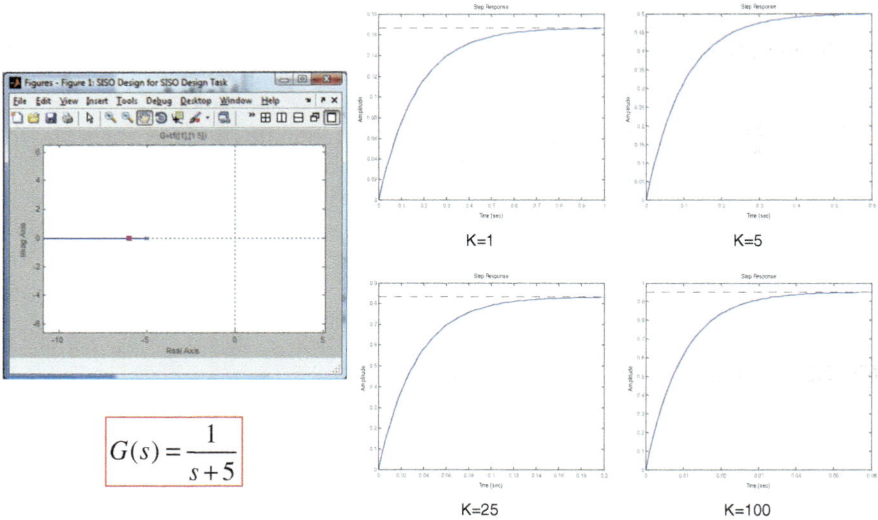

Fig. 4.24 Root Locus for a first order system. This is the simplest case where the root locus covers the real axis in the range [-∞,p_1].

If we now add a second pole, for instance at s=-2, the configuration of the root locus will change, appearing as a rupture point from with the poles can move upward and downward. Thus, according to the value of K, the system will be overdamped, critically damped or underdamped, exhibiting their correspondent behavior (see fig. 4.25)

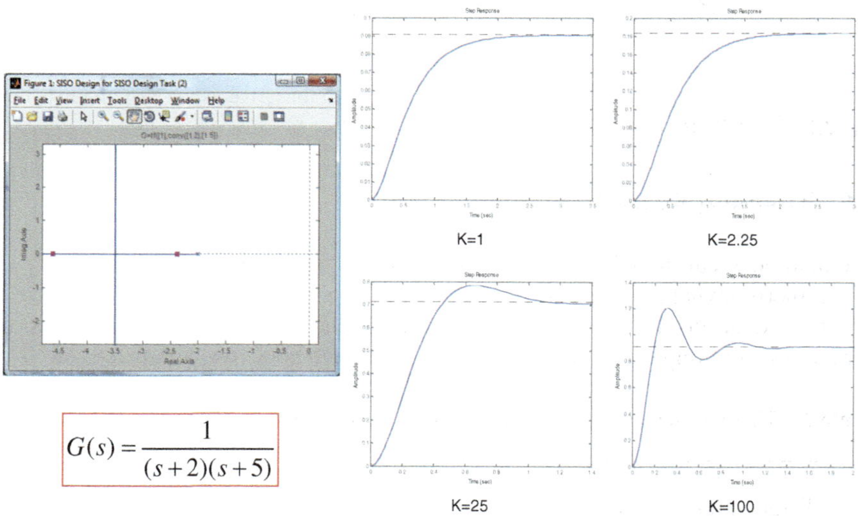

Fig. 4.25 Effect of adding a new pole at s=-2. Note that the root locus turns into a cross making the poles to move over the complex axis. As long as K increases the damping ratio decreases, and thus, the system response oscillations increases.

However, with only two poles the system is stable for any value of K. If we add a third pole to the original system, for instance at s=-7, the root locus bends to the right, existing a value of K that makes the system unstable (see fig. 4.26). Concretely, when K=850, poles are located at s=0.177±8j.

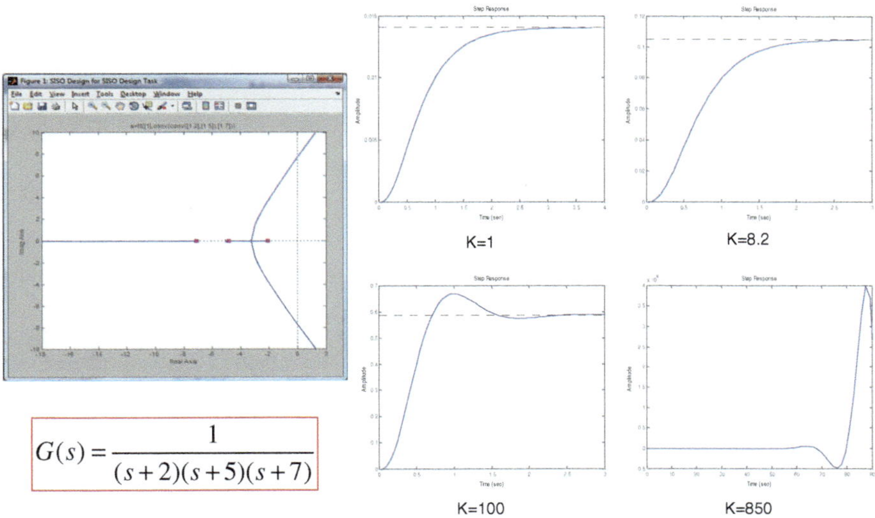

Fig. 4.26 A three-pole root locus example. Adding a new pole bends the root locus to the right. There is a critical value of K that moves the poles to the right part of the s-plane, destabilizing the system.

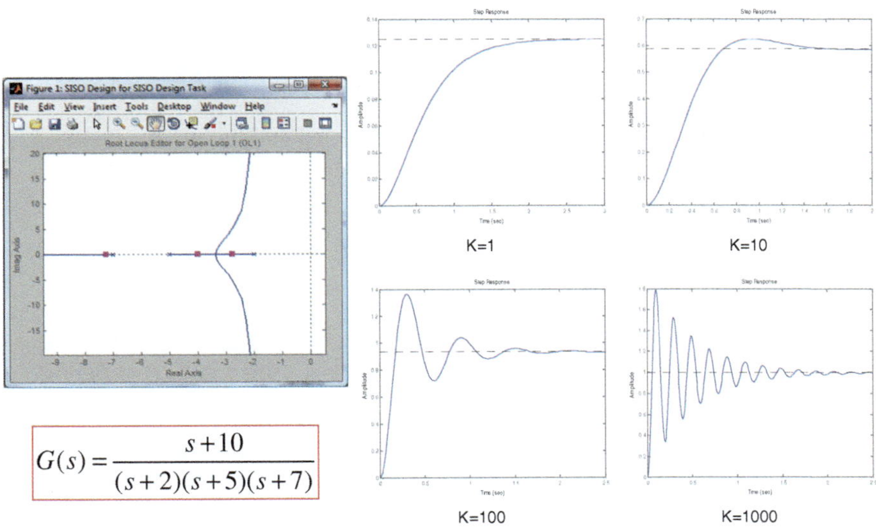

Fig. 4.27 Effect of adding a zero. The addition of a zero at s=-10 corrects the deviation of the root locus, avoiding the destabilization of the system.

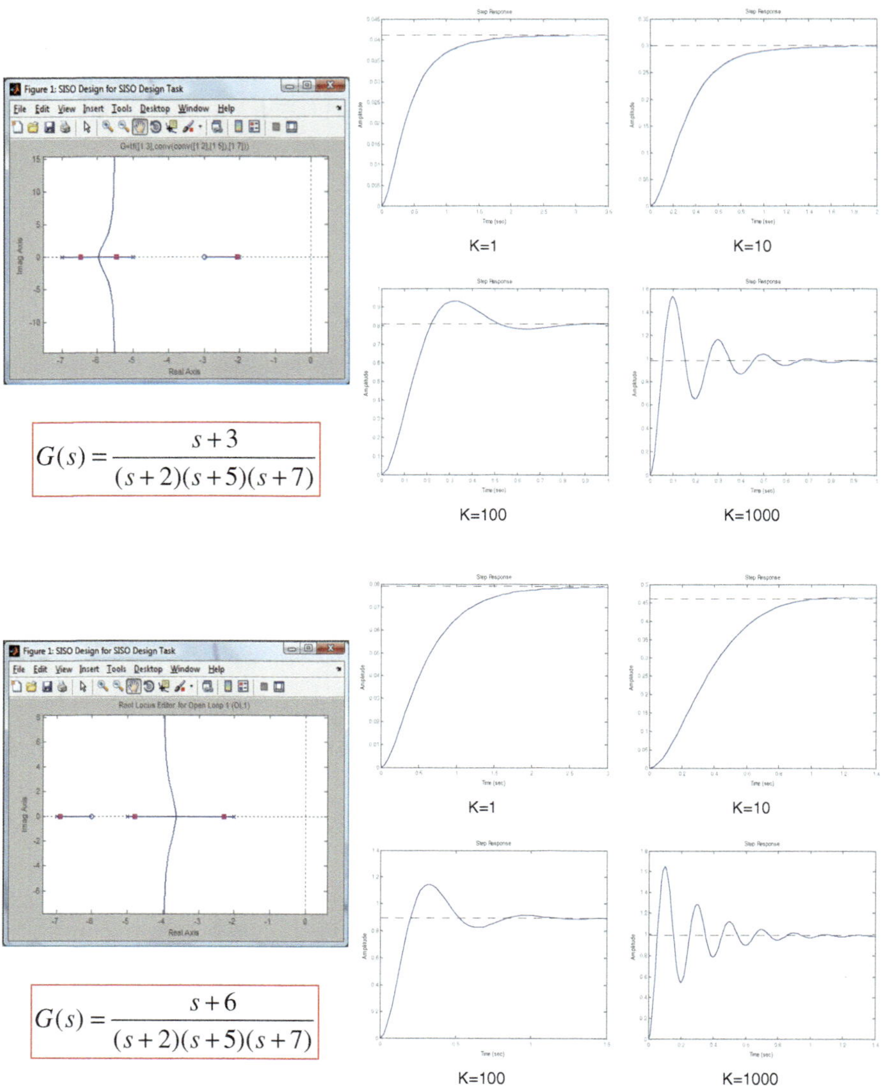

Fig. 4.28 Effect of adding a zero to the three-poles system at different locations. Notice that in any case, the resulting system gains in stability.

Summing up we can realize that increasing the number of poles in the system may lead to instability and undesired effects like overshooting. Contrarily, by adding zeroes to the system we can obtain better results in the system response, we can even stabilize a system. Let's consider now the addition of a zero to the previous example. When adding a zero, e.g. at s=-10, the root locus configuration drastically changes, as shown in fig. 4.27. Now one of the segments ends at the new zero, while the other two exhibit a behavior similar to the case where only two poles are considered. Notice that the system is stable regardless of the value of K, and thus we have introduced stability into the system.

Fig. 4.28 shows the effect of adding one zero at different real locations, proving that in all cases the system keeps stable. The addition of poles and zeroes to a given system is the basis for compensators, i.e. special systems added to the original system to obtain the desired behavior far away from only considering the K parameter. This is explained in detail in chapter 5, where the PID controller is described, which, as an advance here, adds one zero and two poles to the original system.

4.4 Identification

System identification refers to obtaining the transfer function of a system by only considering its response to a given input (see fig. 4.29). This is useful when we have not much information about a system at hand, i.e. it is a black box for us, and we need to find out its transfer function for simulation purposes, for instance.

Although we assume that no information is provided by the system, some hypothesis must to be considered, like for instance the order of the system. In this book we will only focus on system identification of first and second-order systems considering only unitary steps as system inputs.

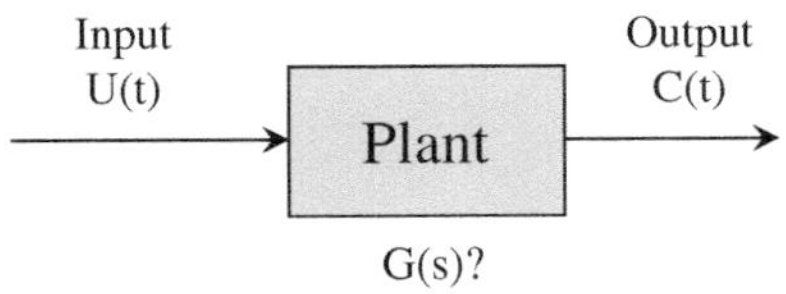

Fig. 4.29 System identification. It consists on finding out the transfer function of the Plant, G(s), by analyzing the system output for a particular input.

4.4.1 First-Order System Identification

As we commented in Section 4.2, a first order system can be represented by a gain, K, and a time constant, τ. Thus, when we try to identify a system assuming it is a first-order system, the goal is to find the values K, and τ.

Example 4.13

Considering a first-order system that when excited with a unitary step input produces the output shown in fig. 4.30. Obtain its transfer function.

Solution

Recalling that parameter K is computed as the relation between the final value of the output and the final value of the input. In the example of the fig. 4.30,

$$K = \frac{y(\infty)}{u(\infty)} = \frac{0.5}{1} = 0.5$$

Remembering that the time constant is defined as the time when the output of the system achieves 63.2% of the final value, by inspection we can conclude that

$$y(\tau) = 0.632 \cdot 0.5 = 0.315 \Rightarrow \tau = 0.5$$

That is, the system whose output is the one shown in fig. 4.30 against a unitary step input is:

$$G(s) = \frac{K}{\tau s + 1} = \frac{0.5}{0.5s + 1} = \frac{1}{s + 2}$$

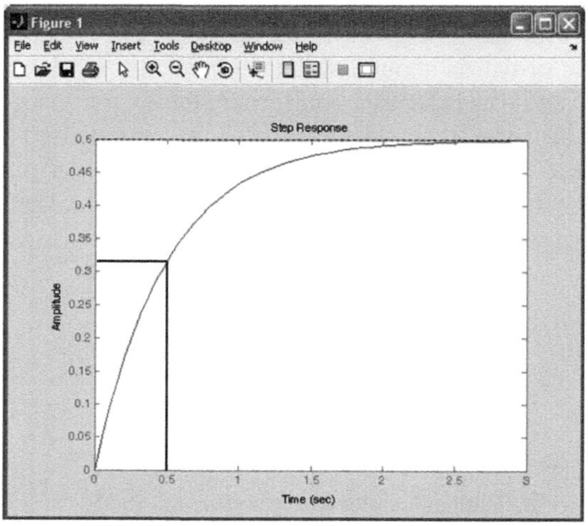

Fig. 4.30 System identification for a first order system. The two parameters to be identified are the system gain and the time constant. The former is calculated as the relation between the final value of the response (in this case 0.5) and the final value of the input. The latter can be obtained by visual inspection, fixing the moment when the response achieves 63.3% of the final value (in this example at t=0.5 seconds).

4.4.2 Second-Order System Identification

The process of system identification becomes more complex when the initial hypotheses indicate that the order of the system is two. Recalling previous sections, a second-order system can be expressed in terms of ξ, and ω_n. According to the damping ratio, ξ, the poles of the characteristic equation can be real and different, real and equal, or conjugate complex, in which case the value of ω_n affects the amplitude of the oscillations. Thus, the behavior of the response against a step input varies in each case, and the aim of identification is to figure out the values of the parameters ξ, and ω_n. To do this, a number of response features are defined whose computation will help us to identify a second-order system.

Having the response of a second-order system when the input signal is a step, it is defined the following (note that some of them have no sense when the system is underdamped or critically damped. Expressions below are approximations and come from analytical manipulations of the general second-order system response function; its in-depth study is out of the scope of this book):

- **Peak Time.** This is the moment when the maximum value of the output is reached. This is an easily measurable parameter which can be mathematically calculated as:

$$t_p = \frac{\pi}{\omega_d} \tag{4.51}$$

where ω_d, is

$$\omega_d = \omega_n \sqrt{1-\xi^2} \tag{4.52}$$

- **Rise Time.** It refers to the needed time to reach the final value for the first time, and thus, sometimes it is called *rise time (0-100)*. Some authors use the *rise time (10-90)* that refers to the time needed to pass from the 10% of the final value to the first time that the response reaches the 90%. The former can be calculated as:

$$t_r = \frac{\pi - \beta}{\omega_d} \tag{4.53}$$

where β=acos(ξ).

- **System overshoot.** The overshoot represents the percentage in which the system output exceeds the final value. Mathematically, it can be expressed as:

$$SO = \frac{y(t_{p)} - y(\infty)}{y(\infty)} \times 100\% \tag{4.54}$$

While the relation of this output characteristic and the parameters of the system is:

$$SO = e^{-\frac{\sigma\pi}{\omega_d}} = e^{-\frac{\xi\pi}{\sqrt{1-\xi^2}}} \tag{4.55}$$

- **Settling Time.** The settling time refers the first time that the system output enters, and is maintained, in a band around the final value. According to the width of such a band, the settling time can be approximated to:

$$t_s = \frac{4}{\sigma}\,(2\%) \qquad t_s = \frac{3}{\sigma}\,(5\%) \tag{4.56}$$

For a 2% and 5% band width of the final value, respectively.

- **Steady State value.** The steady state value is the final value of the system response, i.e. $\lim_{t\to\infty} y(t)$. This final value which can be obtained by mere observation will serve to calculate the gain of the system, as

$$K = \frac{y(\infty)}{u(\infty)} \tag{4.57}$$

Fig. **4.31** depicts most of these characteristics over a underdamped second-order system.

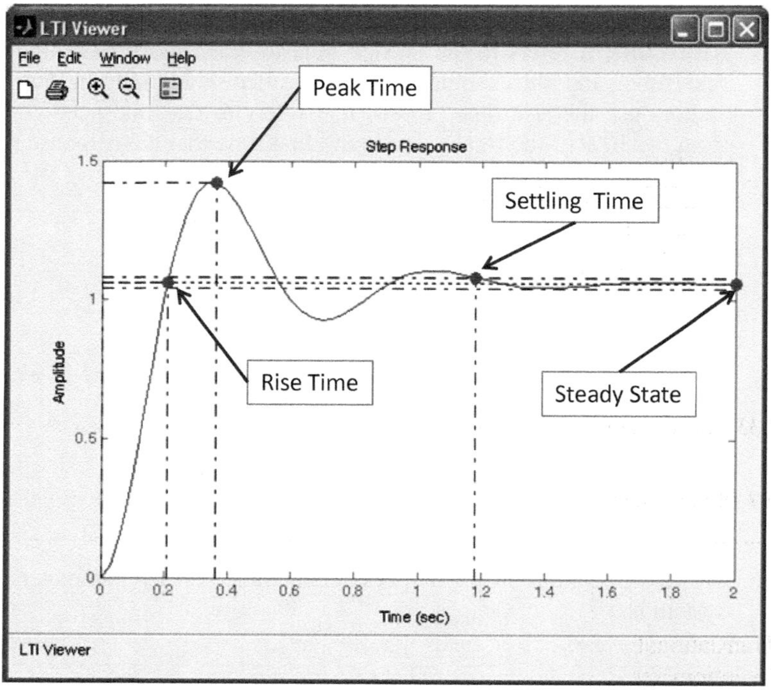

Fig. 4.31 Temporal characteristics of a second-order system visualized through the LTI viewer application.

Let's complete the identification process by calculating the transfer function of the system that causes this output. There is not a unique way to proceed since we have several equations to calculate only two parameters.

Example 4.13

For the system output given in fig. 4.31 and the following information

Peak time=0.364
Overshoot=34.7%
Steady State=1.06

Obtain the transfer function of the system

Solution

From equation (4.51)

$$t_p = \frac{\pi}{\omega_d} \Rightarrow \omega_d = \frac{\pi}{t_p} = \frac{\pi}{0.364} = 8.6307 \tag{4.58}$$

From equation (4.55)

$$\begin{gathered} SO = e^{\frac{-\xi\pi}{\sqrt{1-\xi^2}}} \\ \ln(34.7) = ^{\frac{-\xi\pi}{\sqrt{1-\xi^2}}} \\ \left(\ln(34.7)\right)^2 \left(\sqrt{1-\xi^2}\right)^2 = \left(-\xi\pi\right)^2 \\ \left(\ln(34.7)\right)^2 \xi^2 + \pi^2\xi^2 - \left(\ln(34.7)\right)^2 = 0 \end{gathered} \tag{4.59}$$

$$\xi = 0.3975$$

From equations (4.52), (4.58), and (4.59)

$$\begin{gathered} \omega_d = \omega_n \sqrt{1-\xi^2} \\ \omega_n = \frac{\omega_d}{\sqrt{1-\xi^2}} \\ \omega_n = \frac{8.6307}{0.9176} \end{gathered} \tag{4.60}$$

$$\omega_n = 9.4056$$

Finally the gain of this system is calculated from equation (4.57) and the amplitude of the input as:

$$K = \frac{y(\infty)}{u(\infty)} = \frac{1.06}{1} = 1.06 \tag{4.61}$$

With the values of K, ξ, and ω_n, we can now obtain the transfer function of the system as:

$$G(s) = \frac{K\omega_n^2}{s^2 + 2\xi\omega_n s + \omega_n^2} = \frac{1.06 * 9.4056^2}{s^2 + (2*0.3975*9.4056)s + 9.4056^2} \Rightarrow$$

$$G(s) = \frac{93.77}{s^2 + 70.33s + 88.46}$$

4.5 System Stability

The stability concept refers to a desirable system characteristic for which if (and only if) the input signal is bounded; the output of a stable system will also be bounded. It is important to remark that in this concept the input signal has to be bounded; otherwise, the system output can be unbounded even when it is considered stable.

That is, we are interested in systems that responsed to a certain bounded stimuli, e.g. a step signal with a bounded output. Far from mathematical expressions, the stability concept is widely employed in an uninformed way. For instance, if we try to put a pencil in a vertical position, we rapidly conclude that this is an unstable configuration for this object. In contrast, if we think on a Segway© we can notice that it is stable but only if the driver performs smooth movements; and obviously it will be unstable, i.e. it will fall down, when an abrupt change of the inclination is carried out (and of course, when the control system is switched off).

From a mathematical point of view, we have already referred to system stability when we analyzed the system responses. Remember that in the general case the system response against a step input is:

$$y(t) = \left(K_0 + \sum_{i=1}^{n} K_i e^{-p_i t} + \sum_{j=1}^{r} \frac{K_j}{(n-1)!} t^{n-1} e^{-p_j t} + \sum_{k=1}^{q} K_{k1} e^{-\sigma t} \cos \omega t + K_{k2} e^{-\sigma t} sen \omega t \right) u_e(t) \tag{4.62}$$

Following the definition given at the beginning of this section, the output y(t) has to be bounded and the unique way to assure this requirement is that, for every term,

$$\lim_{t \to \infty} e^{-p_i t} = 0 \tag{4.63}$$

That is, in stable systems $p_i>0$; what is the same, all the roots of the characteristic equation have to lay on the left side of the s-plane. Thus, the most direct way to check the stability of a system is to obtain the characteristic equation and calculate its roots.

Example 4.14

Check whether the following system is stable or not.

$$G(s) = \frac{1}{s^3 + 4s^2 + s - 6}$$

Solution

Computing the roots of the characteristic function we rapidly notice that it is not stable since

```
>> roots([1 4 1 -6])
ans =
   -3.0000
   -2.0000
    1.0000
```

That is, one of the roots is placed on the right part of the s-plane, i.e. there is a pole p1=-1, and therefore one of the exponential terms of the output will not tend to zero. Actually, the step response of this system can be observed in fig. 4.32.

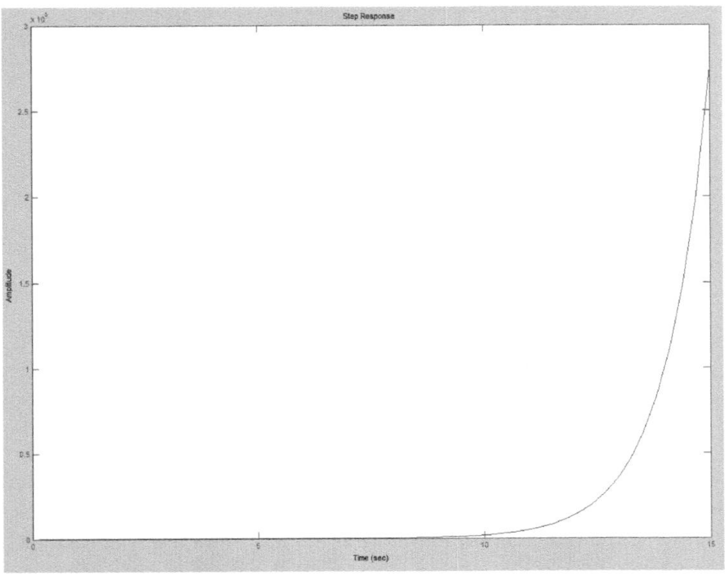

Fig. 4.32 Response of the system G(s) against a unitary step input. Clearly this system is unstable, due to the presence of a pole at the right part of the s-plane.

As you can observe, when studying system stability, sometimes it is not needed to know where the poles are exactly placed, but to know whether all the poles are on the left part of the s-plane. Moreover, if you do not have a computer program (like Matlab) to solve high-order equations, it can be difficult to decide about the stability of a given system. For that, there are some simple rules, given by the *Routh-Hürwitz* theorem, which yields the number of poles that are placed in the "wrong" place, that is, on the right part of the s-plane. The steps to easily check the stability of a system are:

1. Obtain the characteristic equation of the system.

$$a_n s^n + \ldots + a_1 s + a_0 \tag{4.64}$$

2. Assuming that $a_0 \neq 0$, if there is one (or more) term $a_i<0$ when another term $a_j>0$ then one of the roots of the equation, at least, are positive (the system is unstable) or complex conjugated in the form $s=0\pm j\omega$ (critical stability). This condition is necessary but not sufficient.
3. If a characteristic equation passes step 2, there is a possibility of being unstable yet. To completely check the stability of the system we proceed as follows. We construct a table like the one shown in fig. 4.33 where the first two rows are filled up with the coefficients of the characteristic equation: the first row lists the odd coefficients starting from a_n while the second row lists the even coefficients starting from a_{n-1}. The rest of cells of this table are calculated as:

$$t(i,j) = \frac{\left(t(i-1,1) * t(i-2,j+1)\right) - \left(t(i-1,j+1) * t(i-2,1)\right)}{t(i-1,1)} \tag{4.65}$$

a_n	a_{n-2}	a_{n-4}	a_{n-6}	...
a_{n-1}	a_{n-3}	a_{n-5}	a_{n-7}	...
t(3,1)	t(3,2)	t(3,3)	t(3,4)	...
...	...	...	...	...
t(n+1,1)	t(n+1,2)	t(n+1,3)	t(n+1,4)	...

Fig. 4.33 Routh-Hürwitz table. First two rows are filled up with the coefficients of the characteristic equation, while the rest are computed through the expression (4.65).

The Routh-Hürwitz theorem assures that the number of poles located at the right part of the s-plane coincides with the changes of signs of the elements of the first column of the table. Thus, a system is stable if and only such elements are all positive or all negative.

Example 4.15

Check the stability of a system given by:

$$G(s) = \frac{10s+1}{s^4+2s^3+3s^2+4s+5} \tag{4.66}$$

Solution

The characteristic equation for the system (4.66) is

$$s^4+2s^3+3s^2+4s+5=0 \tag{4.67}$$

All the coefficients of (4.67) are positive so we cannot determine the stability of the system yet. We proceed to construct the table (note that empty cells are considered as zero):

1	3	5
2	4	
t(3,1)	t(3,2)	
t(4,1)		
t(5,1)		

being

$$t(3,1) = \frac{(2*3)-(1*4)}{2} = 1 \qquad t(3,2) = \frac{(2*5)-(1*0)}{2} = 10$$

$$t(4,1) = \frac{(1*4)-(2*5)}{1} = -6 \tag{4.68}$$

$$t(5,1) = \frac{(-6*5)-(1*0)}{-6} = 5$$

In this example, notice that the first column of the table (in grey) is 1,2,1,-6, and 5. So there are two changes of sign (from 1 to -6 and from -6 to 5), and therefore the system at hand has 2 poles at the right part of the s-plane: it is unstable.

As expected, solving the characteristic equation we have that:

```
>> roots([1 2 3 4 5])

ans =

   0.2878 + 1.4161i
   0.2878 - 1.4161i
  -1.2878 + 0.8579i
  -1.2878 - 0.8579i
```

Example 4.16

Sometimes, system stability depends on a particular parameter which has to be tuned up to make the system stable. For example given the system:

$$G(s) = \frac{1}{s^3 + 3ks^2 + (k+2)s + 4}$$

Check its stability in function of the parameter *k*.

Solution

We can proceed in a similar way to calculate the range of values that the parameter can take to assure the system stability. The characteristic equation is

$$s^3 + 3ks^2 + (k+2)s + 4 = 0$$

From the second rule of the method we can restrict the value of *k* that makes all the coefficient of the equation to have the same sign as:

$$\left.\begin{array}{l} 3k > 0 \Rightarrow k > 0 \\ k+2 > 0 \Rightarrow k > -2 \end{array}\right\} \Rightarrow k > 0$$

As commented this is only a necessity but not a sufficient condition for stability and we have to complete the process:

1	k+2
3k	4
t(3,1)	
t(4,1)	

$$t(3,1) = \frac{(3k*(k+2)) - (1*4)}{3k} = \frac{3k^2 + 6k - 4}{3k}$$

$$t(4,1) = \frac{t(3,1)*4 - (0*3k}{t(3,1)} = 4$$

Now we have to calculate the range of the value of *k* such as:

$$\left.\begin{array}{l} 3k > 0 \Rightarrow k > 0 \\ \dfrac{3k^2 + 6k - 4}{3k} > 0 \Rightarrow k < -2.5 \wedge k > 0.5275 \end{array}\right\} \Rightarrow k > 0.5275$$

Certainly, when *k=0.5275* the roots of the characteristic equation are:

```
>> roots([1 3*k k+2 +4])
ans =
  -1.5825
   0.0000 + 1.5898i
   0.0000 - 1.5898i
```

and the system is at the limit of the stability. For example for $k=0.5$

```
>> roots([1 3*k k+2 +4])
ans =
  -1.5510
   0.0255 + 1.6057i
   0.0255 - 1.6057i
```

the system becomes unstable, while for greater values, for example $k=0.6$

```
>> roots([1 3*k k+2 +4])
ans =
  -1.6741
  -0.0629 + 1.5445i
  -0.0629 - 1.5445i
```

the system is stable.

4.6 System Stationary Response

In previous sections we were dealing with characteristics of the transient response of systems, paying attention to issues like maximum overshoot, peak times, etc. Although such features are highly relevant for real systems, e.g. it is desirable that a temperature control system responds quickly after a reference signal is given, there is still another relevant parameter which has to do with the stationary response. It can be defined the *steady-state error* of a system as the difference between the reference signal and the stationary system output. This parameter has sense for controlled systems that exhibit a close-loop configuration as shown in fig. 4.34. The steady-state error of systems gives information about the ability of the system to follow an input reference signal. The study and analysis of this parameter is of paramount importance for designing reliable control systems. Let's think, for example, in the altitude control system of an airplane, in charge of keeping an accurate and changing altitude reference given by the pilot; obviously any inherent error in the system on following the pilot instructions may lead to a crash.

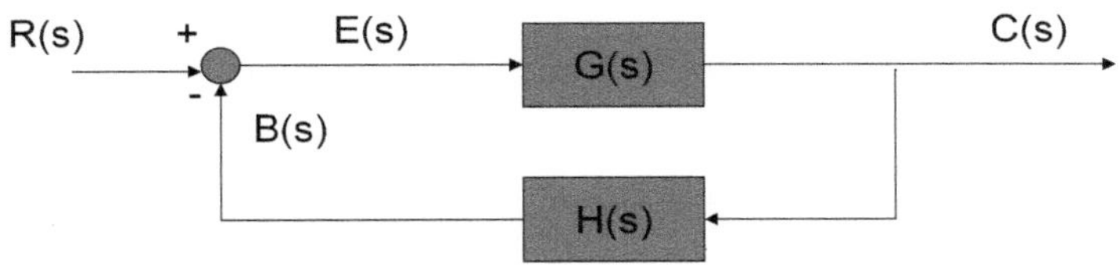

Fig. 4.34 Close-loop configuration considering a sensorial system that measures the plant output. The error is defined as the difference between the reference signal and the measured one.

In the temporal domain, let r(t) be the reference signal and b(t) be system output, probably measured by a sensorial system, the steady-state error can be defined as:

$$e_\infty = \lim_{t\to\infty} e(t) = \lim_{t\to\infty} r(t) - b(t) \tag{4.69}$$

Whereas in the complex domain, the error can be expanded and expressed as shown in equation (4.70).

$$\begin{aligned}
E(s) &= R(s) - B(s) = R(s) - H(s)C(s) \\
E(s) &= R(s) - H(s)\frac{G(s)}{1+G(s)H(s)}R(s) \\
E(s) &= R(s)\left(1 - \frac{H(s)G(s)}{1+G(s)H(s)}\right) \\
E(s) &= R(s)\left(\frac{1+G(s)H(s) - H(s)G(s)}{1+G(s)H(s)}\right) \\
E(s) &= \frac{1}{1+G(s)H(s)}R(s)
\end{aligned} \tag{4.70}$$

Using the equation (4.70) and the final value theorem, we can mathematically define the steady-state error as:

$$e_\infty = \lim_{t\to\infty} e(t) = \lim_{s\to 0} sE(s) = \lim_{s\to 0} \frac{s}{1+G(s)H(s)}R(s) \tag{4.71}$$

Note that the error, obviously depends on the involved systems' nature and also on the characteristics of the system output. That is, regarding the stationary response, a system will behave differently according to the given reference input.

In the following we will particularize the study of the steaty-state error for different input references and types of systems. For that the concept of *type of a system* is considered and defined as the number of integrators of the open-loop system *G(s)H(s)*. That is, rewritting G(s)H(s) as:

$$G(s)H(s) = \frac{K(\lambda_1 s+1)\cdots(\lambda_n s+1)}{s^N(\tau_1 s+1)\cdots(\tau_n s+1)} \tag{4.72}$$

It is said that G(s)H(s) is a *N-type* system. Next, let's check the dependence of the type of a system on the steady-state error when a particular input reference is considered.

4.6.1 Steady-State Error for Step Reference

Equation (4.71), considering *R(s)* as an input reference of amplitude *A*, is rewritten as:

$$e_\infty = \lim_{s\to 0} sE(s) = \lim_{s\to 0} s\frac{A/s}{1+G(s)H(s)} = \lim_{s\to 0}\frac{A}{1+G(s)H(s)} \tag{4.73}$$

Considering $K_p = \lim_{s\to 0} G(s)H(s)$, known as the *position constant*, the steady-state error for step references is defined as:

$$e_\infty = \frac{A}{1+K_p} \tag{4.74}$$

Notice that as long as K_p tends to infinite, the steady-state error tends to zero, and this is only achieved when the system, G(s)H(s), is of type greater than one. That is, from expression (4.72) it can be noted that for $N \geq 1$, the limit will tend to infinite.

Example 4.17

For the open-loop system

$$G(s)H(s) = \frac{2}{s^2+3s+2}$$

Indicate its type and its steady-state error for unitary step input.

Solution

The transfer function can be rewritten as:

$$G(s)H(s) = \frac{2}{s^2+3s+2} = \frac{2}{(s+1)(s+2)} = \frac{2/2}{(s+1)(1/2s+1)}$$

It is a 0-type system and thus the error for a step input at the stationary will not be equal to zero. That is, considering a unitary step input and equation (4.74), the steady-state error is:

$$e_\infty = \frac{1}{1+K_p}, \left\{ K_p = \lim_{s\to 0}\frac{1}{(s+1)\left(\frac{1}{2}s+1\right)} = 1 \right\}$$

$$e_\infty = \frac{1}{1+1} = 0.5$$

4.6.2 Steady-State Error for Ramp Reference

Similarly, when the reference signal *R(s)* is a ramp, equation (4.71) becomes

$$e_{\infty} = \lim_{s \to 0} sE(s) = s\frac{A/s^2}{1+G(s)H(s)} = \lim_{s \to 0}\frac{A}{s+sG(s)H(s)} = \lim_{s \to 0}\frac{A}{sG(s)H(s)} \tag{4.75}$$

Considering $K_v = \lim_{s \to 0} sG(s)H(s)$, called *velocity constant*, the steady-state error for ramp references is defined as:

$$e_{\infty} = \frac{A}{K_v} \tag{4.76}$$

Systems aimed to follow a ramp reference without error at the steady-state must be, at least of type 2 (*N*≥2), in order to make K_v to tend to infinite, and therefore make the error to tend to zero.

Example 4.18

Check the type of the following system and its steady-state error for step and ramp inputs

$$G(s)H(s) = \frac{5}{s^3+3s^2+2s}$$

Solution

Rewriting the transfer function we can check that it is of type 1, i.e. it has an integrator, and thus, it will follow a step input without errors at the steady-state, while it will not accurately respond to a ramp reference.

$$G(s)H(s) = \frac{5}{s^3+3s^2+2s} = \frac{5}{s(s+1)(s+2)} = \frac{5/2}{s(s+1)(1/2s+1)}$$

For a step input of amplitude A, the error will be:

$$e_{\infty(step)} = \frac{A}{1+K_p}, \left\{K_p = \lim_{s \to 0}\frac{5/2}{s(s+1)(1/2s+1)} \to \infty\right\}$$

$$e_{\infty(step)} = 0$$

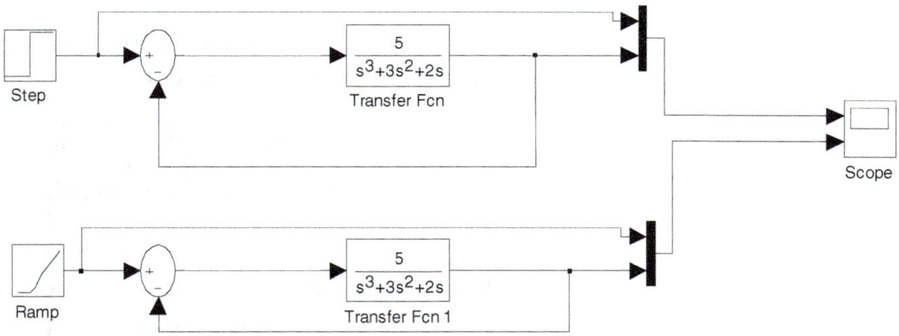

Fig. 4.35 Simulink diagram for step and ramp inputs.

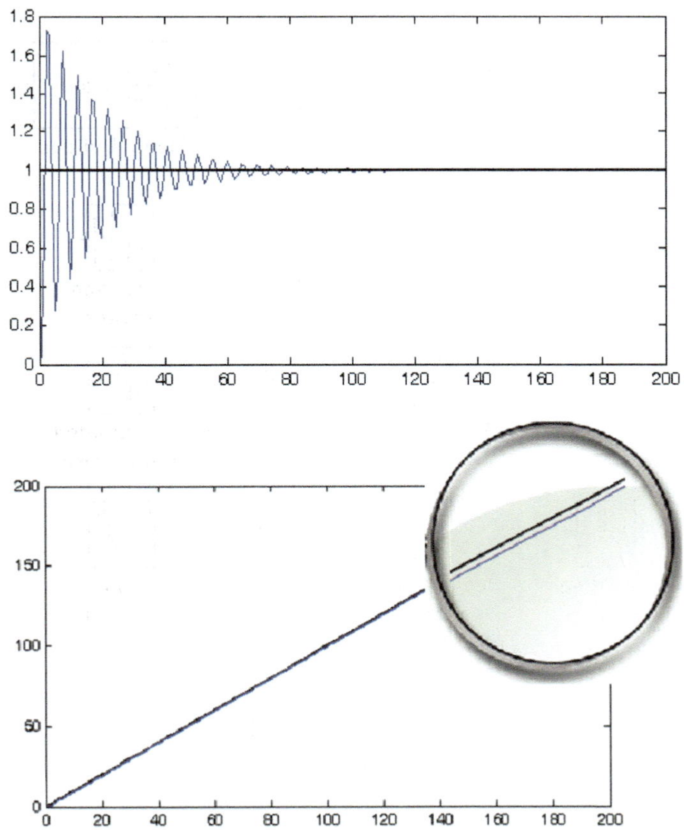

Fig. 4.36 System response. Top) System response for unitary step signal. Given the system is of type 1, the steady-state error is zero. Bottom) System response for unitary ramp signal. In this case the error at the steady state is not zero (it is actually 0.2) as illustrated by the zoomed area.

While for a ramp reference with slope R,

$$e_{\infty(ramp)} = \frac{R}{K_v}, \left\{ K_v = \lim_{s \to 0} s \frac{5/2}{s(s+1)(1/2s+1)} = \frac{5}{2} \right\}$$

$$e_{\infty(ramp)} = \frac{R}{\frac{5}{2}} = \frac{2R}{5}$$

As expected the system has no errors at the steady-state for steps references but has a certain error for ramps since it is a 1-type system (see fig. 4.36 and fig. 4.36).

The same reasoning can be done for other input references, for instance, for parabolic references, $R(s)=1/s^3$, the error will be calculated as:

$$e_\infty = \lim_{s \to 0} sE(s) = s \frac{A/s^3}{1+G(s)H(s)} = \lim_{s \to 0} \frac{A}{s^2 + s^2 G(s)H(s)} = \lim_{s \to 0} \frac{A}{s^2 G(s)H(s)} \tag{4.77}$$

And therefore the type of a system aimed to follow this input without error should be of type 3 or higher.

Chapter 5
Introduction to Control Systems

Control systems are aimed to modify the behavior of an existing system to perform in a desired way. Several examples can be found in the real life in which certain control actions are needed to achieve the wanted results. For example the temperature of a room may change due to external and unexpected perturbations, e.g. somebody opens/closes a window, sunshine heats the walls, etc. An air-conditioned system can be installed to help in keeping a given temperature in spite of any external change in the system. The air-conditioned system includes elements to modify the current temperature (also called *actuators*, since they act on the system), *sensors* to measure the current temperature, and some *control* logic to determine the actuation to be carried out in order to achieve the target temperature. In this chapter we will introduce the basis of the control systems, their influence on real life systems, and their design to modify the behavior of a given plant to fulfill particular specifications.

5.1 Control Systems

Control systems have been present in the society from the ancient past in an attempt to design and construct machinery with a particular purpose. For instance, the first records of the oldest control system date from the III century BC, when *Ktesibios* designed a water clock, knows as *Clepsydra*, based on a mechanism in charge of maintaining constant the water inflow of a tank. By means of a floating valve the water volume of an auxiliary tank is maintained invariable and thus the inflow of water of the controlled tank constant. The water level of the latter indicates how much time is passing.

However, the real expansion and formal study of the control systems came into scene at the end of the XVIII century with the arrival of the Industrial Revolution. Representative examples of this period are the *self-regulating wind machine* devised by E. Lee in 1745 which was aimed to automatically control the orientation and inclination of the windmill's sails according to the wind direction improving the mill performance. Another example is the *steam engine centrifugal governor* envisaged by Watt in 1788 to regulate the speed of steam locomotives (see fig. 5.1).

Nowadays, we can find control systems almost everywhere, from highly sophisticated and complex systems to maintain the speed and course of planes, to simple and cheap ones in charge of maintaining the temperature of an electric water heater.

With independence of their complexity, we can identify the following elements in any control system:

1. The desired output, also known as *reference*. For instance, the desired temperature in a typical temperature control system.
2. The *plant*, that is, the system whose output is controlled.
3. A *sensorial system* in charge of measuring the plant output.
4. An *acting system* that directly acts on the plant according to the inputs given by the controller.
5. *Perturbations*, that is, external and unforeseeable variations of the control system that can affect to both the output of the plant, e.g. opening a door in a temperature control system, and to the sensor measurements, e.g. a misleading sensor reading due to, for instance, electric interferences.
6. The *controller* responsible for calculating the proper inputs for the acting system *(control actions)* according to the difference between the reference and the plant output *(error)*, in spite of perturbations.

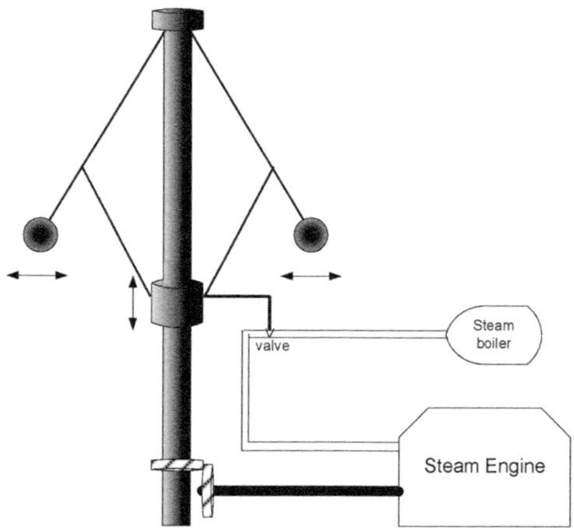

Fig. 5.1 The steam engine centrifugal governor. The governor regulates the steam valve by means of the centrifuge force caused by the locomotive speed.

The configuration of the components of a control system gives designers two different options: open-loop and close-loop. fig. 5.2 depicts a scheme of the different components of a control system in a *close-loop* configuration. This is the most extended and useful configuration, probably also the most obvious way, to control the output of a given plant since the continuous feedback provided by the sensorial system permits the controller to update its actions[1]. However *open-loop* configuration (see fig. 5.3) can be also found in certain simple and cheap control systems. In an open-loop configuration no feedback is supplied (note that the sensorial system disappears) and the controller gropes for achieving the reference. Obviously, this type of control systems cannot manage plant perturbations and are only suitable for simple, cheap, and deterministic systems, e.g. the controller for managing the speed of the head of a CD burning unit is set to a given, fixed, voltage according to the selected burning speed.

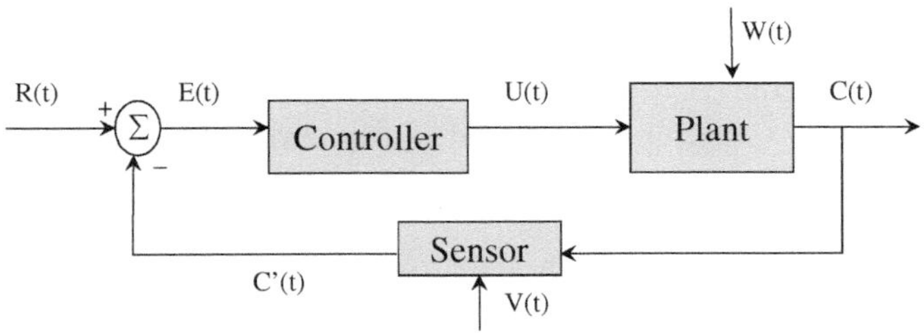

Fig. 5.2 Close-loop control configuration. Perturbations in the plant and in the sensor are continuously managed by the controller, enabling the system to properly react.

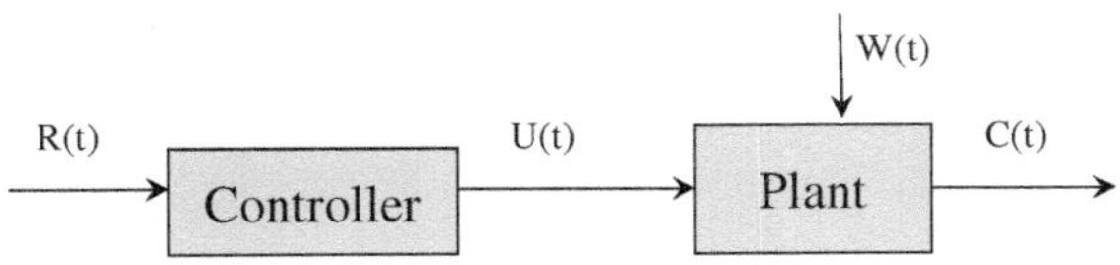

Fig. 5.3 Open-loop control configuration. The controller provides a fix and pre-programmed control action based on the reference. Any perturbation on the plant will not be managed which would lead to catastrophic situations.

When designing, the complexity and accuracy of the considered control system should be coherent with respect to the characteristics of the application at hand. Thus, in the temperature control example, the control system may be as simple as considering a close-loop configuration in which the controller switches on/off the air-conditioned machine at a fixed power when the room temperature is higher/lower than a given threshold. This solution would be enough and cheaply implemented for the temperature control application in an apartment, for instance, but would be insufficient, even catastrophic for solving the same problem applied to a different context, e.g. controlling the temperature of the computing room of a bank, or a nuclear plant. That is, according to the special necessities and requirements, the control system to be designed should fulfill given specifications

[1] The time required for completing a loop is critical and largely depends on the application at hand. For example for controlling the temperature of an apartment, a couple of seconds (even minutes) can be enough, but for controlling the speed or position of an aircraft, it should be closed in milliseconds.

that normally are related to a) the *swiftness* to achieve the desired output (or at least a bounded approximation to it), and b) the *accuracy* of the obtained output with respect to the commanded reference.

Next sections describe different control strategies based on non-linear systems, like *relays*, and on *PID* controllers.

5.2 Basic Control Actions

A naïve, but at the same time effective, way to control certain systems is by means of the so-called *bang-bang control*. This strategy consists of providing two different and prefixed control actions according to the measured error. Typically it is implemented by switching on/off and actuator when the error is higher/lower than a given threshold (for this, the bang-bang control is also known as on/off control).

On/off control is normally considered for slow and not critical systems, for instance in electric braziers or water heaters in which the temperature is controlled by a bimetallic thermostatic switch that disconnects the power when a limit temperature is achieved and its internal metal components split up. When temperature goes down, the metal components of the thermostatic switch contract and they contact, switching on the brazier again.

Although on/off control can be effective, the response of the controlled system exhibits severe inconveniences. The plant output slowly approaches the reference, and what is worse, it may approach in an oscillatory way. The following example illustrates these effects.

The SIMULINK diagram shown in fig. 5.4 represents a close-loop control system aimed to produce a plant response equal to the reference value '1'.

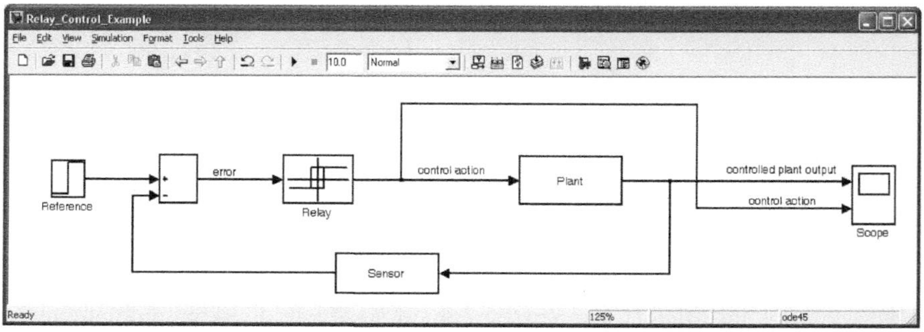

Fig. 5.4 Closed-loop control system based on a relay. The controller provides two fixed control actions depending on the error and on pre-established thresholds.

The control logic has been designed with a relay block configured to produce the control action '1' when the error (the difference between the plant response and the reference) is positive, i.e. greater than *eps*, and '0' when the error is negative, i.e. less than *eps*[2]. This behavior can be configured through the relay configuration dialog shown in fig. 5.5.

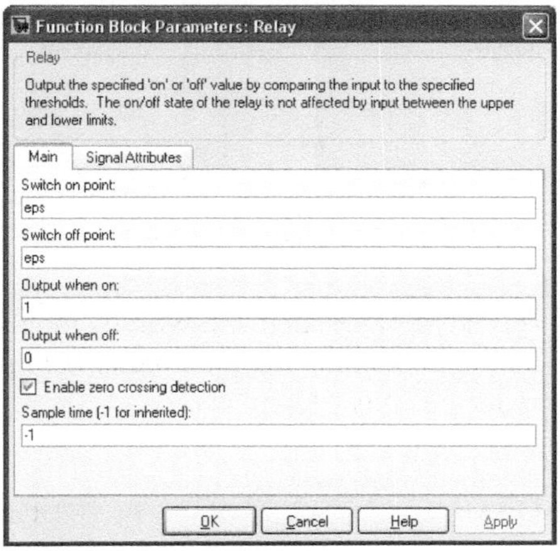

Fig. 5.5 Relay block parameters. A fix control action (output when on/off) is produced according to the input of this block and the switch on/off points. Hysteresis can be implemented by increasing the margin between the switch on and switch off points.

Example 5.1

Study the effect of bang-bang control on the plant defined by the transfer function:

$$G(s) = \frac{5}{s^2 + 2s + 1}$$

and a sensorial system assumed to be G(s)=1.

Solution

Notice that the plant presents a gain of '5', that is, the plant response given a unit reference will go to '5', in open-loop configuration and in absence of a controller. Considering an on/off controller configured as shown in fig. 5.5 will produce a controlled response that tends to the reference. Such a controlled response is shown in fig. 5.6a, while the control action is plotted in fig. 5.6b. Note that the response is slow and exhibits a significant overshoot. This is due to slow reactivity of the on/off controllers. Also note that every time the relay switches, the response oscillates, and as long as the error is being minimized, the relay switches more frequently and the amplitude of the response oscillations decrease. The oscillation effect of the response is not clearly visible in fig. 5.6a, but observe it in fig. 5.6c where the last 2 seconds of the response have been zoomed in.

[2] The value of the constant *eps* in MATLAB is 2.2204e-016.

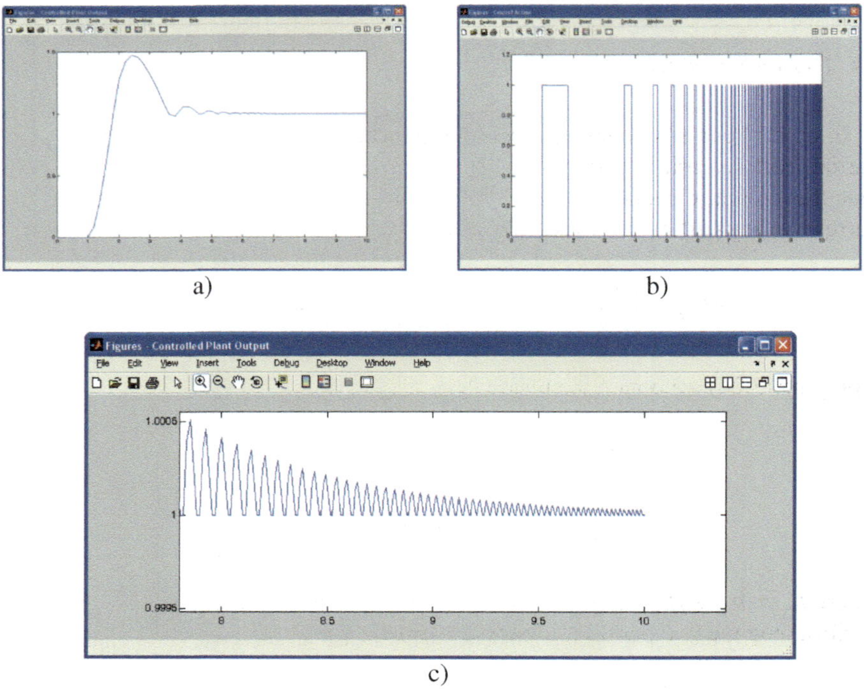

Fig. 5.6 Outputs from the example system. a) Controlled plant response. b) Control action. Notice that as long as the output achieves the reference, the frequency of the relay switches increases. c) Steady-state controlled response zoomed in. Oscillations are due to the continuous changes in the control action.

The performance of on/off control systems is largely tight to the amplitude of the *hysteresis* area defined. In this example, the hysteresis occurs in the interval (*-eps,eps*) which ensures an inappreciable oscillation of the controlled response at the steady state, but causes a high number of relay's switches as observed in fig. 5.6c. From a theoretical point of view a reduced interval for changing the control action is acceptable, even ideal relays can be considered where the switch is produced at a crisp threshold, i.e. the switch on and off points are equal. However, from a practical point of view, the implementation of a control system through relays with a narrow hysteresis area produces the wear of the employed components due to the continuous activation/deactivation, reducing their operation life. On the other hand, considering a higher hysteresis area would bear in a higher steady time and oscillations in the response.

Summing up, the on/off control provides cheap solutions to non-critical systems where non-precise responses are assumable.

5.3 PID Controller

The PID controller is one of the most widespread control techniques in engineering and has being used since the 1930's. PID stands for *Proportional,*

Integral, and *Derivative* since this type of controller acts proportionally to the error, to the integral of the error, and to the derivative of the error. The idea behind the PID control is that acting proportionally to the error will produce a prompt response of the controlled output, acting proportionally to the integral of the error will eliminate the error at the steady state and finally, being proportional to the derivative of the error will reduce oscillations.

Mathematically, the PID control action can be defined in the time domain as:

$$u(t) = K_p\left(e(t) + \frac{1}{T_i}\int_0^t e(t)dt + T_d\frac{de(t)}{dt}\right) \tag{5.1}$$

which is normally rewritten as:

$$u(t) = K_p e(t) + K_i\int_0^t e(t)dt + K_d\frac{de(t)}{dt},$$
$$K_i = \frac{K_p}{T_i} \quad K_d = K_p T_d \tag{5.2}$$

With this, the transfer function of a PID is as follows (see fig. 5.7):

$$G_{PID}(s) = \frac{U(s)}{E(s)} = K_p + \frac{K_i}{s} + K_d s = \frac{K_d s^2 + K_p s + K_i}{s} \tag{5.3}$$

That is, a PID controller adds two zeros and one pole to the plant.

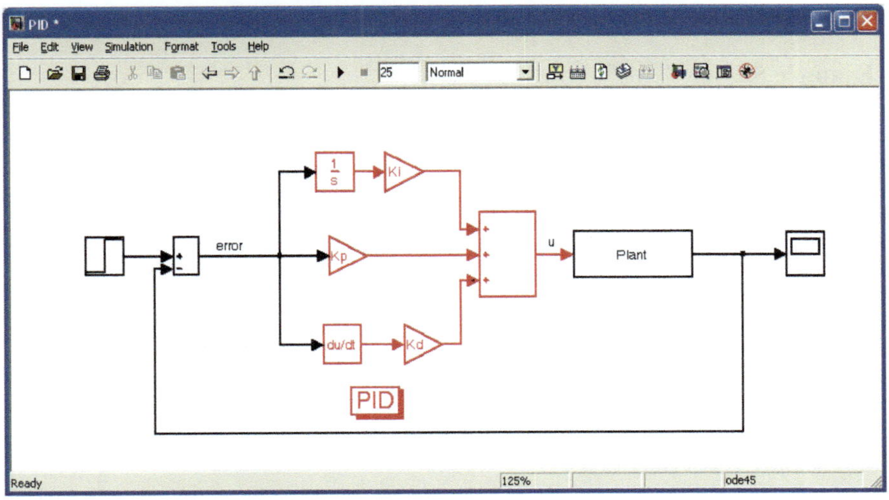

Fig. 5.7 The PID control. The PID control, shown in red, acts over the plant proportionally to the error, its integral and its derivative.

Next, the components of the PID control and their effects are individually described in more detail.

5.3.1 Proportional Control

The proportional control relies on acting proportionally to the measured error. That is, the higher the magnitude of the error is, the stronger the control action is applied to the plant. This is an elemental and intuitive control action that improves the plant output but also may lead to undesirable effects.

Actually human reactions usually follow this type of control. Let's think for example you are driving your car, trying to be at the center of the lane. This is actually a control system where your brain is the controller, your hands directly act on the vehicle (the plant) through the steering wheel, your eyes are the sensors, and the error is the distance of the car to the center of the lane. Small deviations are corrected via small actuations, however an extremely high error caused, for example, by a sudden gust could make you react excessively, swerving to keep the track in a impulsive way that may lead the car to swerve from side to side, requiring additional actuations on the steering wheel. The control action is directly proportional to the instant error. Small errors cause slight corrections, but a sudden excessive error causes an instinctive one and totally out of proportion. In the worst case the system will turn into unstable, i.e. a traffic accident in this example.

The proportional controller (see fig. 5.8) is configured through the parameter, K_p, which indicates how strong the control action is according to the current error.

Example 5.2

Consider a proportional control of a system represented by the transfer function:

$$G(s) = \frac{5}{s^3 + 9s^2 + 20s + 10}$$

Check the time response for different values of $K_{p.}$

Solution

The gain of G is 5/10=1/2, and thus, the step response of the system in open-loop will achieve that value.

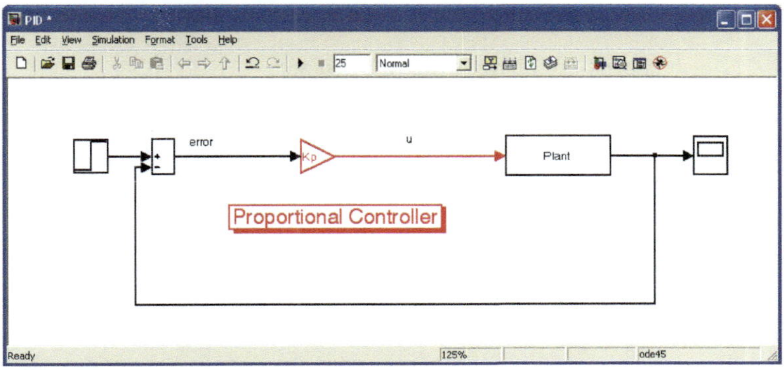

Fig. 5.8 The Proportional Control. The control action, u, is proportional to the error, (u=error*K_p), enabling the system to quickly react against perturbations and/or changes in the reference, but at the expense of produce oscillations, even instability, when the error or K_p are sufficiently high.

The closed control loop, considering a proportional controller is represented by the transfer function:

$$G_{CL}(s) = \frac{5K_p}{s^3 + 9s^2 + 20s + 10 + 5K_p}$$

Note that when K_p=1 the final value of the response is 5/15=0.333. As long as K_p becomes higher, the error at the steady state decreases. In theory when $K_p \to \infty$, the error at the steady state tends to 0. However, increasing the value of K_p produces oscillations in the output response when the magnitude of the error is considerably high. At an extreme situation, the system may become unstable given a certain value of K_p. fig. 5.9 shows the plant output for different values of K_p where considering a value of K_p=25 yields a final value of 125/135=0.926, but at the expense of extremely high oscillations.

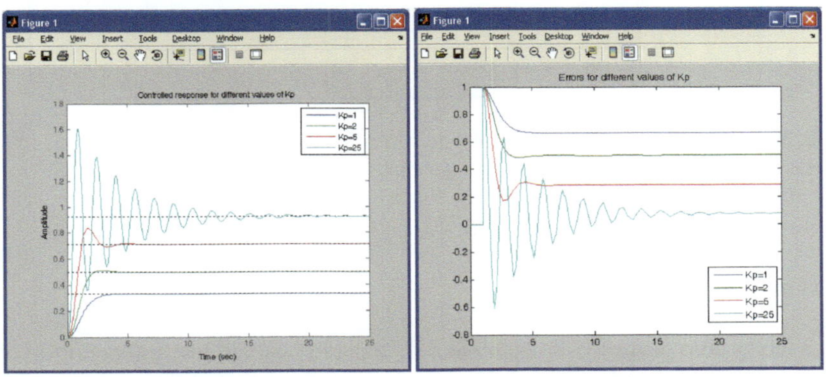

Fig. 5.9 Influence of the K_p parameter in the Proportional Control. As long as K_p increases, the system reacts more impulsively against the error, reducing the error at the steady-state but causing more oscillations.

Summing up, the *pros* of a proportional controller are:

- It speeds up the response of the plant.
- It improves the accuracy, i.e. reduces the final error.

The *cons* of a proportional controller:

- It increases the oscillations of the output.
- It increases the steady-state time.
- It may lead to system instability.

5.3.2 *Integral Control*

Considering only the *instant* error to control the plant is not a good option to achieve a highly accurate and smooth response in plant output. The proportional control does have not records of past experiences, i.e., of the evolution of the response along the time, and thus its exhibits the aforementioned limitations.

An improvement on the proportional control is the addition of the *integral control*, obtaining the *PI* controller, as shown in fig. 5.10. A control action proportional to the integral of the error is added to eliminate the error at the steady-state.

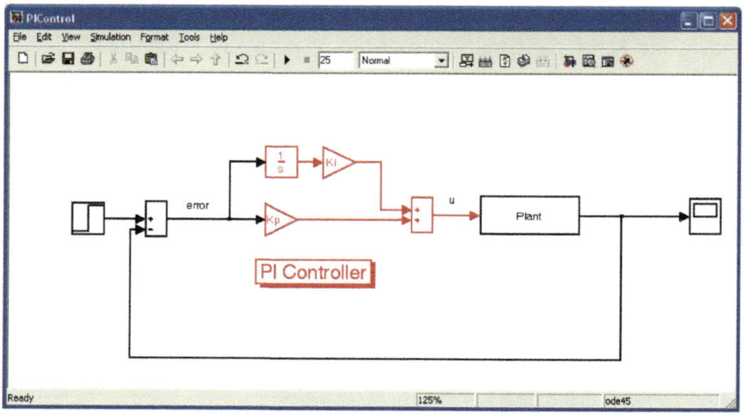

Fig. 5.10 The PI control. The control action considers now both the error and its integral. Taking into account the integral of the error permits this controller to eliminate the error at the steady state.

Example 5.3

For the system proposed in example 5.2, consider also an integral action over the error, and check its effect for different values of K_i.

Solution

The closed-loop transfer function is now:

$$G_{CL}(s) = feedback(series\left(K_p + \frac{K_i}{s}, \frac{5}{s^3 + 9s^2 + 20s + 10}\right), 1)$$

$$G_{CL}(s) = \frac{5K_p s + 5K_i}{s^4 + 9s^3 + 20s^3 + 10s + 5K_p s + 5K_i}$$

That is, the system gain is 1, i.e. $\frac{5K_i}{5K_i}$.

The integral component of the PI adds an integrator, and thus the *type* of the system increases, eliminating the error at the steady-state when the input reference is a *step*.

Let's observe this effect considering the previous example with Kp=5 and Ki=1 (see fig. 5.11). Note that the system response (blue line) has slightly increased the overshoot but, most importantly, the error at the steady-state has been completely eliminated.

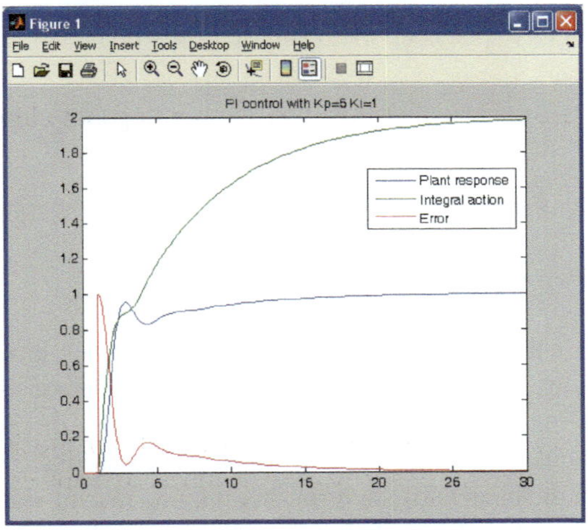

Fig. 5.11 The PI Control. After the first peak, the response tends to the reference. Note how the error (red line) tends to zero, while at the steady state, the input to the plant is only produced by the integral component.

The effect of varying the parameter K_i reduces the steady-state time and the settling time at the expense of largely increasing the overshoot of the system as shown in fig. 5.12. Note how the red response reach its settling time at time t=~10 sec. (and the blue one around 20), but the former exhibits a 40% of overshoot.

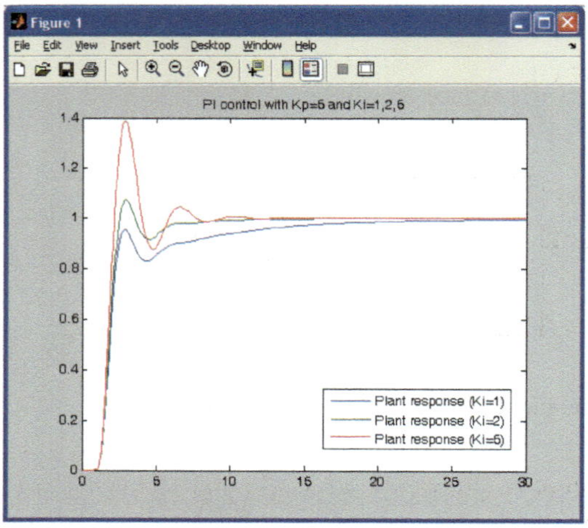

Fig. 5.12 The PI Control. Notice that increasing the integral parameter the output achieves the steady state faster but at the expense of a high overshoot.

Therefore, the *pros* of a integral controller are:

- It largely enhances the accuracy, totally eliminating the error at the steady state.
- It slightly speeds up the response of the plant and reduces the settling time.

The *cons* of a integral controller:

- It increases the overshoot and oscillations of the output.

5.3.3 Derivative Control

Finally, the third component of the PID controller comes into scene to reduce response oscillations. The two other components have demonstrated their ability to produce a fast and precise response, but at the expense of causing oscillations and high overshoots. This is because in both cases the controller has no information about the change of the error, but only its instant value and its integral along the time. The derivative of the error provides valuable information about how the error value is going and acts to establish it on a constant value. The derivative control will act firmly when there are strong changes in the error, typically during the slopes of the error.

Example 5.4

Repeat the previous example including a derivative action with parameters K_p=5, K_i=5 and K_d=1. Try different values for Kd and observe the effects on the system output.

Solution

Fig. 5.13 depicts the system output for the example considered in the previous section with K_p=5 and K_i=5 (the situation illustrated by the red line of the fig. 5.12.) adding a derivative action with K_d=1. Notice that including a derivative control, an additional control action opposes the impulsive action of the proportional and integrative components. This occurs when the derivative of the error is negative, which coincides with the overshoot.

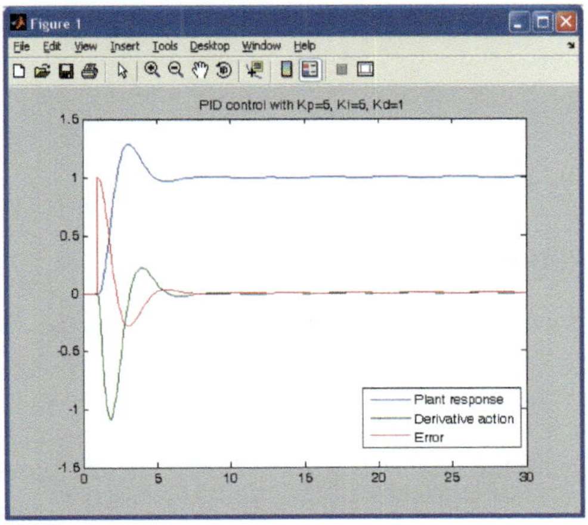

Fig. 5.13 The PID control. Adding a derivative control the response is smoothed, reducing the overshoot and oscillations[3].

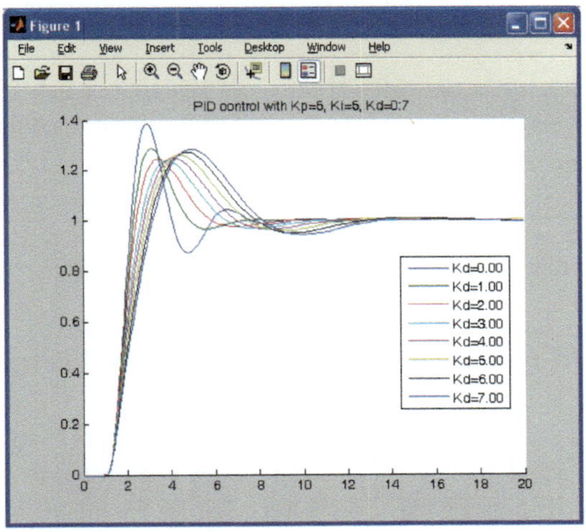

Fig. 5.14 The PID control. The derivative control strives for reducing the overshoot, however, high values of the parameter K_d may cause the opposed effect, even make the system unstable.

---

[3] To produce this plot, the derivative action has been limited to be properly shown since when the step is raised, the derivative of the error tends to $-\infty$.

Varying the value of K_d will cause a slower and smoother response, where the response oscillations are flattened. See fig. 5.14 for values of K_d up to 3 (the cyan line). However note that higher values for this parameter will have the opposed effect, it causes an increment in the overshoot. This is because as long as the parameter increases, the control action is stronger and stronger, producing an excessive reaction to the controlled variable, i.e. the derivative of the error.

Thus, the *pros* of a derivative controller are:

- It reduces the overshoot and oscillations in the plant response.

The *cons* of a PI:

- It slows down the output.

5.3.4 PID Summary

As stated in the previous sections, each component of the PID control exhibits valuable advantages but, also has undesirable disadvantages. Table shown in fig. 5.15 summarizes the contribution of the different parameters of a PID. Note that the effects are sometimes conflictive since for instance, trying to reduce the rise time by increasing the value of K_p, will lead to an increment in the overshoot, which could be solved by modifying the value of K_d, that in turn, could modify the rise time. That is, the selection of the parameters of a PID controller is a tradeoff between the response characteristics and it is responsibility of the designer, based on his experience, to tune it accordingly.

	Rise Time	**Overshoot**	**Setting time**	**Error**
K_p	Diminishes	Increases	Slight Change	Diminishes
K_i	Diminishes	Increases	Diminishes	Disappears
K_d	Slight Change	Diminishes	Diminishes	Slight Change

Fig. 5.15 Contribution of each component of the PID to the system response.

5.4 Design of Controllers: Ziegler-Nichols Techniques

In this section we focus on the design of PID controllers, that is, on the tuning of the parameters K_p, K_d, and K_i to meet given requirements. Although different mechanisms for designing PID controllers can be found in the literature, most of them are based on experimental and specific techniques. In this book we will only focus on the Ziegler-Nichols techniques.

Ziegler and Nichols described in 1942 two simple recipes for helping engineers to properly tune PID controllers. These methods are completely experimental and do not require to model exactly the system, albeit, some assumptions about them have to be done. They broadly involve considering the system as a black box that yields a certain output when it is excited with a given input. By analyzing such outputs, the Ziegler-Nichols techniques advise the gain values for the proportional, integrative and derivative components of the PID. It is important to highlight that the methods described in the following are intended to be general enough to be applied to a variety and diversity of systems and thus the results are not always as good as the desired ones. Thus, the solutions provided by the Ziegler-Nichols should be taken as a starting point in the tricky chore of PID tuning, which should be properly tuned by hand to achieve the required output characteristics.

5.4.1 Ziegler-Nichols Technique in Open Loop

The Ziegler-Nichols technique in open loop, also known as first method of Ziegler-Nichols, can be applied when the output of a system, when unit step is considered has a '*S*' shape. That is, it is applicable when the system response resembles a first order output with, probably, a small delay. Systems containing integrators or dominant complex poles will not exhibit this behavior and thus, this recipe is not applicable for them.

The steps to tune a PID controller following this method are:

1. Consider the system in open loop and excite it with a step input of amplitude A (typically A=1)
2. Approximate the response to the typical first-order system output, identifying the time constant, τ, the response delay L, and the system gain K.
3. Use these values to calculate the gains for each controller component through the following table.

Table 5.1 PID gains provided by the first method of Ziegler-Nichols

Controller	Kp	Ti	Td
P	$\frac{\tau}{KL}$	-	-
PI	$\frac{0.9\tau}{KL}$	$\frac{L}{0.3}$	-
PID	$\frac{1.2\tau}{KL}$	$2L$	$0.5L$

Example 5.5

Design a PID controller for a plant whose step response in open-loop is as shown in fig. 5.16.

Solution

Note that the system response exhibits the '*S*' shape needed for applying this method. We consider a first-order system modeled by a time delay *L* and a time constant τ. These parameters can be calculated by tracing the tangent to the output at the inflection point. Crossings with the x-axis and the final value determine the required parameters. Note that this computation of τ is only valid for pure first-order system. Since we are assuming an arbitrary system as a first-order one, it would be more precise to rely on the formal definition of the time constant, as the moment in which the output achieves 63.3% of the final value.

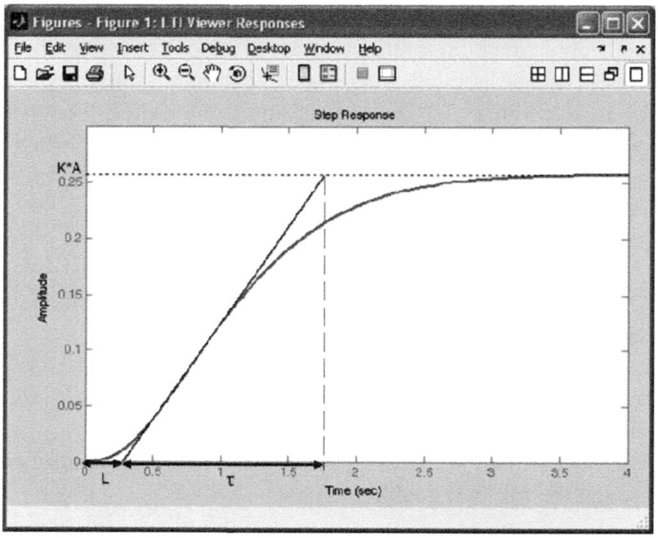

Fig. 5.16 Obtaining the parameters for the open-loop Zigler-Nichols Method.

In this example, *L=0.2754*, τ=1.7605-L=1.4851, and K=0.26. With these values, a PID controller tuned through this method will be:

$$\mathbf{K_p} = 1.2\frac{\tau}{KL} = 1.2\frac{1.4851}{0.26 * 0.2754} = \mathbf{20.74}$$

$$T_i = 2L = 0.558, \left\{K_i = \frac{K_p}{T_i}\right\} \Rightarrow \mathbf{K_i} = \frac{20.74}{0.558} = \mathbf{37.1693}$$

$$T_d = 0.5L = 0.1377, \left\{K_d = K_p T_d\right\} \Rightarrow \mathbf{K_d} = 20.74 \cdot 0.1377 = \mathbf{2.856}$$

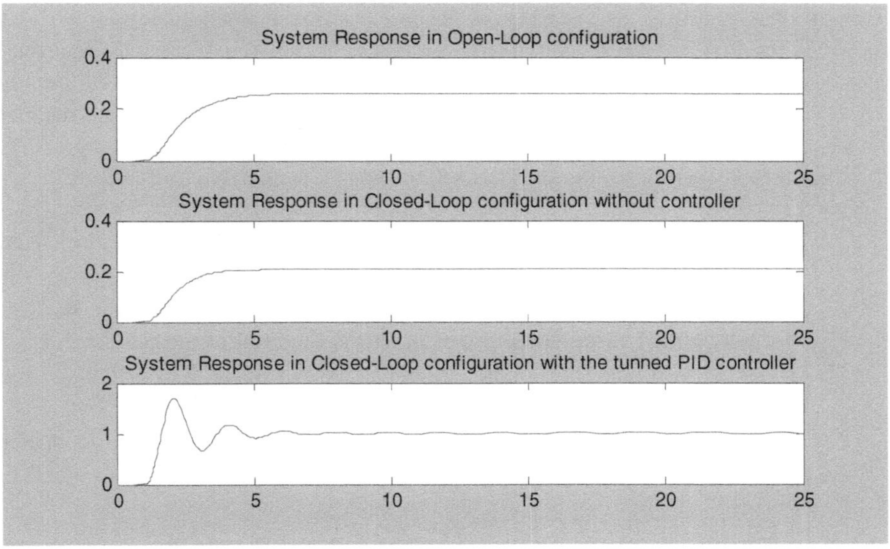

Fig. 5.17 System responses. Up) Open-loop output, used for obtaining the parameters L and τ. Middle) Closed-loop without controller, i.e. a proportional controller with K_p=1. Bottom) PID controlled response. Note that the output, achieves the reference, although it exhibits a remarkable overshoot.

Fig. 5.17 illustrates the benefits of using the tuned PID. While the non-controlled system presents a very slow response with a significant error at the steady state, when the obtained PID controller is considered, the output eliminates such an error (what is normal because of the integral component), at the expense of some oscillations and a maximum overshoot over 50%. Although the use of the tuned PID controller enhances the system response, it may be desirable to improve it even more. For that, and starting with the values obtained through the Ziegler-Nichols method, a manual tuning should be necessary. Please refer to the end of the next section where some considerations when manually adjusting PID controllers are outlined.

A final consideration should be taking into account when using this method. The prerequisite of approximating the system response to a first-order response with a delay makes us to calculate the time constant τ through its geometrical definition, i.e. tracing the tangent line at the inflexion point. However it is only valid for (real) first-order systems. In order to eliminate this problem, it is recommendable to consider that τ is defined as the moment where the output achieves the 0.63% of its final value by simple inspection of the response.

5.4.2 Ziegler-Nichols Technique in Close Loop

The second recipe for tuning PID controllers, known as the *close-loop technique* can be only applied to systems which can be turned into unstable under proportional control. This method was designed to obtain a diminishing overshoot between the first and the second oscillation in a 25%. Obviously such a result may not be assumable in most of the cases, especially when the maximum oscillation is excessive, and thus this method is normally used as a preliminary step to a manual tune.

The steps to be followed in this second method of Ziegler-Nichols are:

1. Consider only proportional control, as shown in fig. 5.18, and provide a unitary step as reference.
2. Increase K_p upto a certain value, the so-called critical value K_{cr}, for which the controlled output oscillates at the limit of the stability.
3. Analyze the system output to assess the period of the oscillation, P_{cr}, when considering the critical proportional gain K_{cr}.
4. Use the values of P_{cr} and K_{cr}, to calculate the gains for each controller component through the following table.

Table 5.2 PID gains provided by the second method of Ziegler-Nichols

Controller	Kp	Ti	Td
P	$0.5K_{cr}$	-	-
PI	$0.45K_{cr}$	$\frac{P_{cr}}{1.2}$	-
PID	$0.6K_{cr}$	$\frac{P_{cr}}{2}$	$\frac{P_{cr}}{8}$

Example 5.6

Consider a plant modeled by the following transfer function:

$$G_{PLANT}(s) = \frac{40}{2s^3 + 10s^2 + 82s + 10}$$

Design a PID controller according to the second method of Ziegler-Nichols.

Solution

Notice that, for applying the recipe for this method it is not mandatory to have this information about the plant, but it has been considered here for clarity sake.

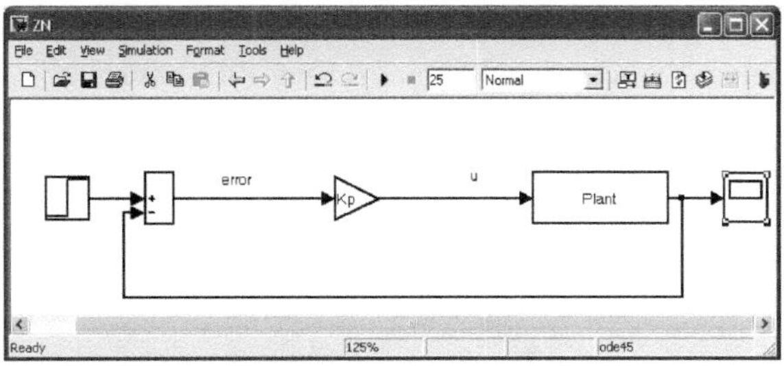

Fig. 5.18 Simulink diagram of the configuration for applying the second method of the Ziegler-Nichols tuning technique.

Steps 1) and 2) are carried out by considering a proportional control of the considered plant (see fig. 5.18), for which, the value of Kp is being increased experimentally until reaching a oscillatory but stable output. This is achieved in our example for a value $K_p=K_{cr}=10$, for which the output is depicted in fig. 5.19.

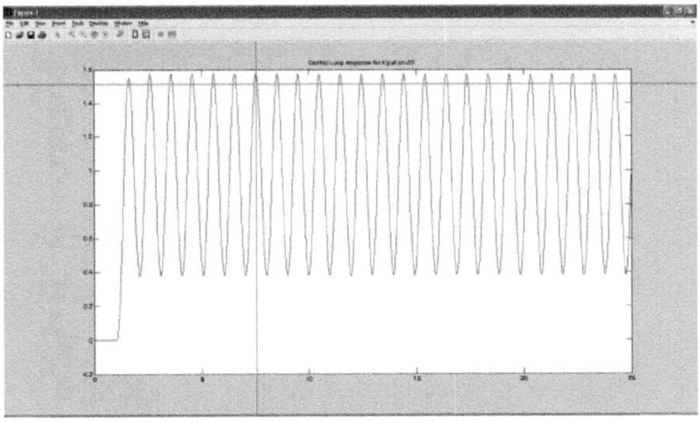

Fig. 5.19 Maintained oscillations caused by setting a proportional control that leads the system to the stability limit.

The step 3) requires the calculation of the period of the output oscillations. Such a period can be manually measured through the Matlab plot through the command *ginput*.

```
>> [x,y]=ginput(2)
x =
     6.5524
     7.5605
y =
     1.5669
     1.5784
```

The above command permits us to click over the figure twice, yielding the exact clicked position (*x*, y). We can obtain the critical period P_{cr} by clicking on two consecutive maximums of the oscillation and subtract the *x* values to obtain a positive period. In this case, P_{cr} = 7.5605-6.5524=1.0081 s.

Having, K_{cr} and P_{cr}, the parameters needed for this recipe, the step 4) can be performed to obtain the K_p, K_i and K_d gains for a PID controller:

$$\boldsymbol{K_p} = 0.6K_{cr} = 0.6 \cdot 10 = \mathbf{6}$$

$$T_i = \frac{P_{cr}}{2} = \frac{1}{2} = 0.5, \left\{ K_i = \frac{K_p}{T_i} \right\} \Rightarrow \boldsymbol{K_i} = \frac{6}{0.5} = \mathbf{12}$$

$$T_d = \frac{P_{cr}}{8} = \frac{1}{8} = 0.125, \left\{ K_d = K_p T_d \right\} \Rightarrow \boldsymbol{K_d} = 6 \cdot 0.125 = \mathbf{0.75}$$

The application of the closed-loop technique of Ziegler-Nichols has been illustrated in an experimental fashion, which is the usual case when no information about the plant is at our disposal. However, in certain situations we can also calculate the K_{cr} and P_{cr} parameters mathematically. The reasons are obvious; exact solutions without the need of acting on the system (which may be dangerous when trying to achieve an oscillatory response). On the contrary, as commented previously, this option is not always feasible since it requires the exact model of the plant (or at least a good approximation). Next, having the transfer function of the plant, the mathematics behind this example is outlined.

The characteristic equation for the closed loop when a proportional control is considered is:

$$2s^3 + 10s^2 + 82s + 10 + 40K_p = 0$$

The evaluation of this equation at point $s = 0 \pm j\omega$ yields the value of the critical proportional gain, $K_p = K_{cr}$, that moves the poles of the system to the $j\omega$ axis, causing, thus, a maintained oscillation at angular speed ω.

$$2s^3 + 10s^2 + 82s + 10 + 40K_p \Big|_{s=0\pm j\omega} = 0$$

$$2(j\omega)^3 + 10(j\omega)^2 + 82j\omega + 10 + 40K_p = 0$$

$$-2j\omega^3 - 10\omega^2 + 82j\omega + 10 + 40K_p = 0$$

$$-10\omega^2 + 10 + 40K_p = 0 \tag{1}$$

$$-2j\omega^3 + 82j\omega = 0 \tag{2}$$

From (2) we can find the value of ω:

$$\omega\left(-2\omega^2 + 82\right) = 0 \Rightarrow$$

$$\omega = 0,$$

$$\omega^2 = \frac{-82}{-2} = \sqrt{41}$$

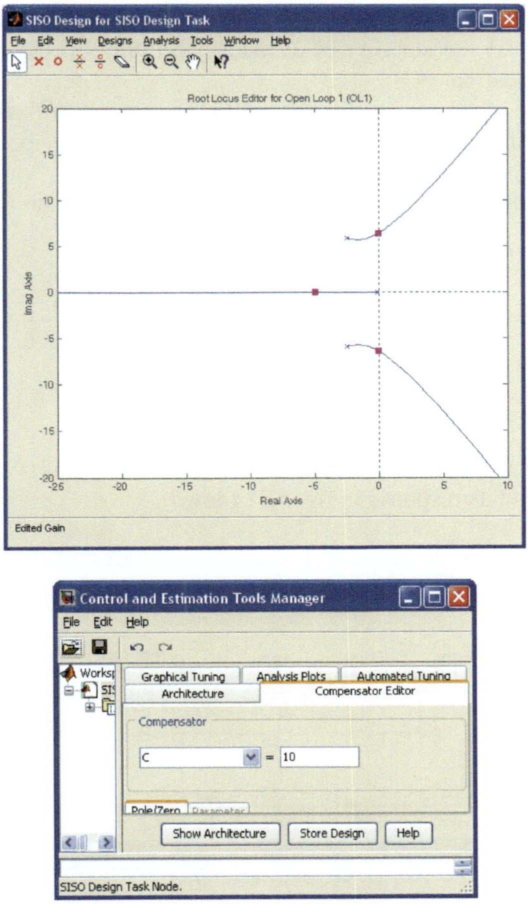

Fig. 5.20 Root-locus of the plant. Observe that the value for Kp=10 (called compensator in rltool) moves two of the poles to $j\omega$ axis, i.e. to the limit of the stability.

From (1) and using the results[4] of (2) the value of K_p is found:

$$K_p = \frac{10\omega^2 - 10}{40} \Rightarrow$$

$$(\omega = \sqrt{41}) \Rightarrow K_p \frac{400}{40} = 10$$

That is, K_{cr}=10, and the critical period P_{cr} is calculated as $P_{cr} = \frac{2\pi}{\omega} = \frac{2\pi}{\sqrt{41}} = 0.9813$ which reveals our manual precision when using the *ginput* command.

4 Note that ω =0 has not been used since the proportional gain is defined for our purposes in the range [0..∞].

The value of K_{cr} could have also been calculated by means of the Rooth criterion, that tests the stability of a system according to the coefficients of the characteristic equation.

Depicting the root locus of the system through the *rltool* (see fig. 5.20) we can confirm the position of the poles when the proportional gain, called in this tool *compensator*, C, is set to 10 as well as the value of ω.

Finally, with these new values K_{cr}=10 and P_{cr}=0.9813, the gain values for a PID are:

$$\boldsymbol{K_p} = 0.6K_{cr} = 0.6\cdot 10 = \mathbf{6}$$

$$T_i = \frac{P_{cr}}{2} = \frac{0.9813}{2} = 0.4906, \left\{K_i = \frac{K_p}{T_i}\right\} \Rightarrow \boldsymbol{K_i} = \frac{6}{0.4906} = \mathbf{12.2287}$$

$$T_d = \frac{P_{cr}}{8} = \frac{0.9813}{8} = 0.1227, \left\{K_d = K_p T_d\right\} \Rightarrow \boldsymbol{K_d} = 6\cdot 0.1227 = \mathbf{0.736}$$

With these gains, the SIMULINK for the final controlled system is shown in fig. 5.21 (up), and its response in fig. 5.21 (down).

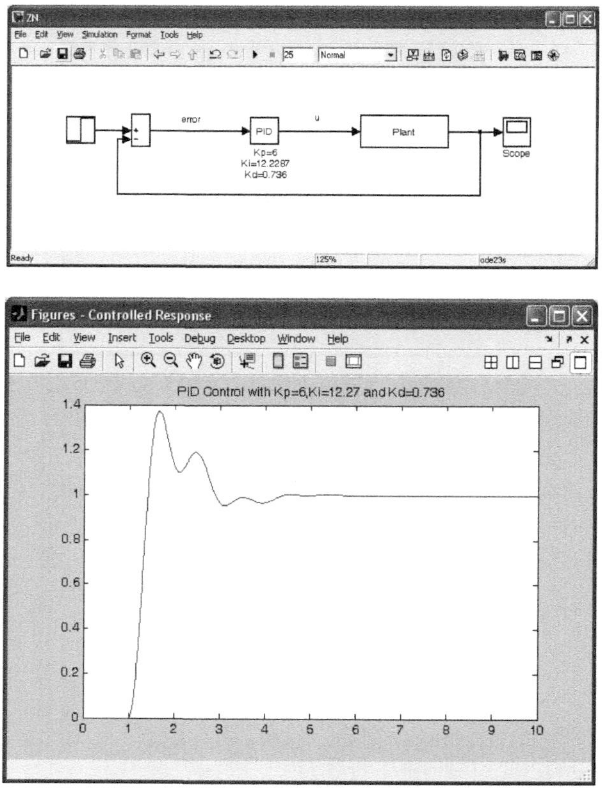

Fig. 5.21 The tuned PID controller. Up) the SIMULINK diagram for the final system. Down) the controlled system output. Note that swiftness is achieved at the expense of a considerable overshoot.

Notice that the controlled output exhibits a considerable overshoot, although its second oscillation fulfills the 25% diminishing rule. Also note the reduced raising time, mainly caused by the effect of the Kp and Ki gains, and also the work of the integral component, eliminating the error at the steady state.

In order to evaluate the response characteristics of this output accurately, we rely on the *ltiview* tool. Previously, we had to obtain the LTI object for the controlled system since ltiview cannot deal directly with the SIMULINK diagram. For that, we construct the transfer functions for the tuned PID and the plant, and perform the series and feedback operations. The followed sequence of commands and the result of the ltiview execution are shown in fig. 5.22.

From the results of the ltiview tool we derive that:

- The system response is considerably fast with a rising time $t_r = 0.273$.
- It exhibits notable oscillations with a remarkable maximum overshoot of 36.5%.
- The steady state is achieved shortly, with a settling time $t_s = 3.17$ and a null error because of the contribution of the integrative component.

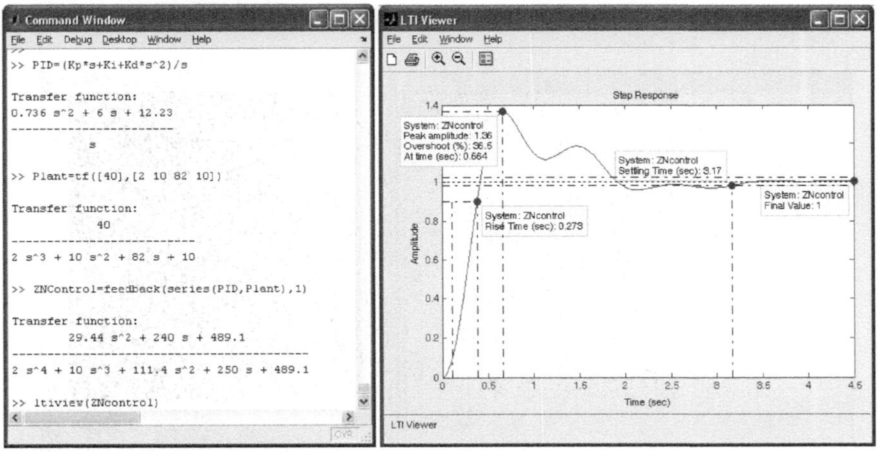

Fig. 5.22 Response characteristics for the tuned PID control through ltiview. Left) Computation of the LTI object through the transfer functions of the tuned PID, the plant, and feedback and series operations. Right) Main window of the ltiview tool. Note how the excessive overshoot (36.5%) is rapidly reduced giving a settling time of around 3.2 seconds.

Although this result may be valid for certain control systems, for example for controlling a domestic water heater, it would result inadmissible for other more critical situations requiring higher precision. Think for example that the plant considered in this example represents the servomechanism of the steering system of the vehicle. The reference input is given by the orientation of the steering wheel and the output of the system is the orientation of the vehicle's wheels. For instance considering the current tuned PID for this example, when the driver turns the steering wheel 20º, the vehicle's wheel will achieve such a value after 0.5 seconds approximately (the raising time from 0-100). After that the vehicle will be oriented to ~27º and then will oscillate until the final value is the desired one. You will agree, this situation is not funny when driving at 100 Km/h.

Being patent the weakness of the tuning provided by the Ziegler-Nichols method, we rely now on a manual adjustment, taking into account the provided values and the desired output. For example we will try to adjust such values to get a quick response, minimizing oscillations.

Manual adjustment is complex since the effects of components are interconnected and affect the output in opposite ways. We recall here the table of the fig. 5.15 with the effects of each component of the PID over the characteristics of the system response that reveals the cross effects of each component.

The key idea in this example is to reduce the overshoot by reducing the values of K_p and K_i, which in turn will increase the rise time. One can think that increasing K_d sufficiently would fix the problem, but this is not the solution given that in certain cases it produces more oscillations, although with lower amplitudes. We have tried different options and one of the best is considering K_p=5, K_i=2 and K_d=1, assuring a relatively fast response with small oscillations and without error at the steady state (see fig. 5.23).

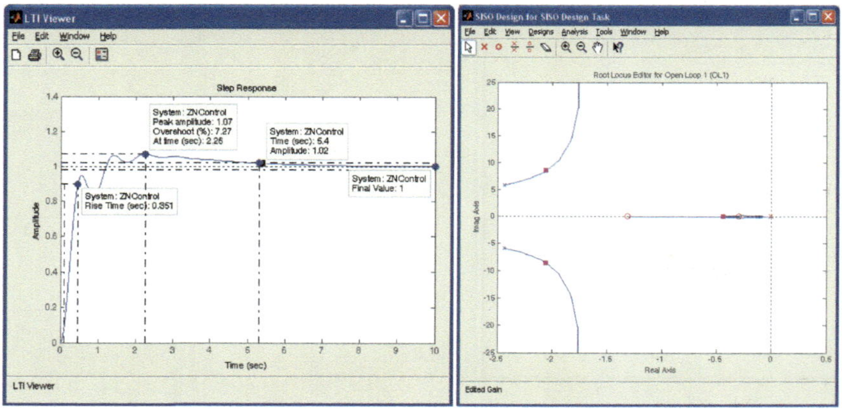

Fig. 5.23 Results of the PID manual adjustment. Left) system response. Note that overshoot has been considerably reduced. Right) Root locus of the controlled system. The presence of complex poles in the close-loop contributes with oscillations in the output. The zeros added by the PID and the proportional gain has been carefully selected to obtain a quick but not too much oscillating response.

It is remarkable that the nature of the considered system does not allow us to improve much more the response by only considering a PID. The plant exhibits three poles (two of them are complex) and no zeros. The PID controller adds, as previously commented, one integrator a two real zeros, resulting in a fourth order system with two zeros. In this situation the controlled system will always exhibit complex closed-loop poles (as shown in fig. 5.23, and thus, the output will oscillate.

5.5 Applications

In this section we will deal with the use of rltool for adjusting a PID controller through an illustrative example. Let's consider a plant represented by the transfer function:

$$G(s)=\frac{6}{2s^2+3s+1} \tag{5.4}$$

For which, the following restrictions are established: a) Maximum overshoot < 5%, b) Settling time< 2 seconds, and c) null steady-state error for step inputs. Factorizing the denominator of expression (5.2), it can be noticed that system is of type 0, and thus, it presents a steady-error against step input. As commented, a PID controller adds an integrator to the system, and thus, it increases the type of the system, which would solve the requirement c).

$$G(s)=\frac{6}{(2s+1)(s+1)} \tag{5.5}$$

A PID controller also adds two zeroes to the open-loop system whose location will affect, primarily, the transient response of the controlled system. Thus, in order to fulfill the proposed constrictions, we can proceed by placing two zeroes within the root locus provided by rltool, adjusting their position until the constrictions are met.

Fig. 5.24 shows the root locus of the original system, where the correspondent constraints have been considered. The yellow area of the s-plane indicates the incorrect position of the poles in close-loop for achieving the restrictions of maximum overshoot and established settling time. To fulfill such restrictions we should be able to move the poles of the system out of the yellow area, which is impossible for the current system.

When adding an integrator, i.e. a pole at s=0, the root locus changes notably, as shown in fig. 5.25. Now there is a segment that starts at the pole at s=-1 and runs tending to -∞ over the real axis. Thus there will be a value for the compensator K that moves out this pole from the yellow area, but not the others. Moreover, the root locus is bended to the right, making the system unstable for certain values of the compensator. Concretely this system is unstable for $K \in [0.25,\infty)$.

Fig. 5.26 depicts the step response for the system for the marked position of poles in the close-loop. Note that the settling time restriction is far from being met and that, it is not fulfilled regardless of the value of the compensator.

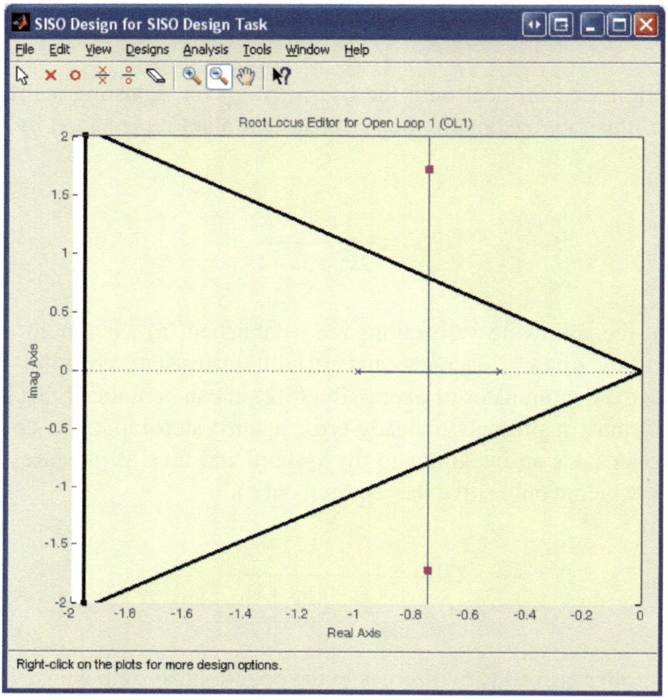

Fig. 5.24 Root locus for the second-order system considered in the example. Note that in this configuration poles will be always located within the yellow area, and thus the two restrictions cannot be simultaneously met (the restriction a), i.e maximum overshoot <5%, is satisfied for $K \in [0, 0.227]$.

To fully design a PID controller, there remains the addition of two zeroes to the system. There is no a unique solution and it depends on the ability and expertise of the designer. In this case we have placed one zero at s=-0.2, relatively close to the integrator. Thus the segment of that starts as the integrator goes over the real axis until it reaches the added zero. The other zero has been placed at s=-2, modifying the aspect of the root locus as shown in fig. 5.27. Notice that in this last configuration all the root locus runs on the left part of the s-plane, and thus, the system will be stable regardless of the value of the compensator K. Also note that two out of the three poles can be moved out to the white area which represents the allowed zone where the poles of the system must be able to fulfill the two imposed restrictions. It is important to highlight that the zones displayed by rltool cannot be directly applicable, since they are defined in terms of second-order systems. In our case as

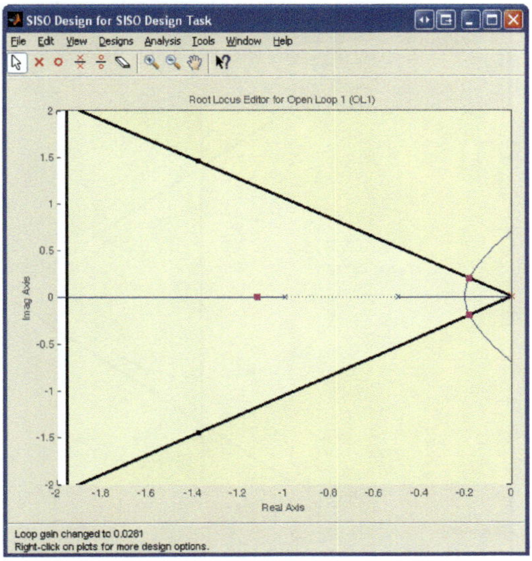

Fig. 5.25 Root locus after adding an integrator. The addition of a pole makes the system more unstable. In this configuration there are values of the compensator that moves some poles to the right part of the s-plane.

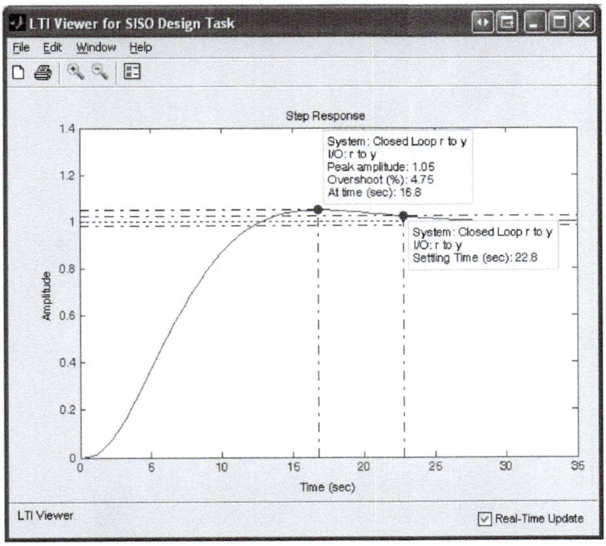

Fig. 5.26 System response. Although the maximum overshoot restriction is achieved, the settling time is far from the desired 2 seconds. In this configuration increasing the compensator K will make the system unstable.

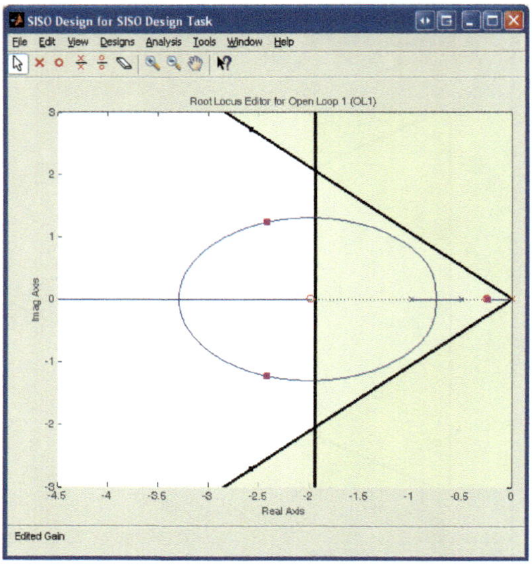

Fig. 5.27 Root locus after adding the integrator and two zeroes. Note that the addition of zeroes makes the system to gain in stability. Close-loop poles are shown for K=0.6. Notice that, for this value of the compensator, the integrator has been cancelled with the added zero at s=-0.2.

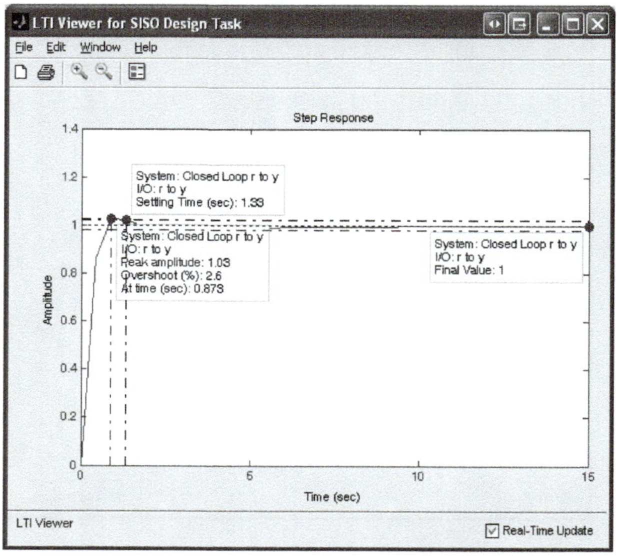

Fig. 5.28 Controlled response. The three constraints have been achieved: overshoot < 5% (2.6%), settling time < 2 seconds (1.33 seconds) and no steady-state error against unitary step (final value is 1).

long as the compensator increases, the segment that started at the integrator rapidly approach the added zero at s=-0.2. When this occurs, it is said that the pole (integrator) and the zero are *canceled*, and the system can be approximated to a second order system, being the constraint areas displayed in that case representative of the constraints to be checked.

Fig. 5.28 illustrates the response of the controlled system for K=0.6. Settling time has been reduced to 1.33 seconds and the maximum overshoot to 2.6%.

Once the requirements have been fulfilled, the transfer function of the designed controller is shown by rltool (see fig. 5.29).

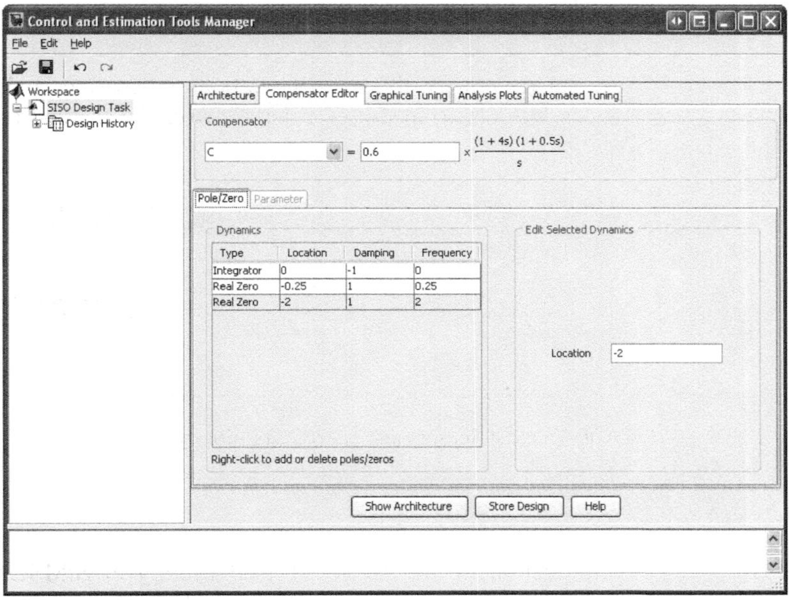

Fig. 5.29 Designed controller. Rltool constructs the transfer function of the additional poles and zeroes included in the root locus.

For this example, the designed controller can be represented by a transfer function with an integrator, two zeroes and a certain gain, i.e. it is a PID controller, given by

$$G(s) = \frac{0.6 \cdot (1+4s)(1+0.5s)}{s} = \frac{1.2s^2 + 2.7s + 0.6}{s} \tag{5.6}$$

Identifying expression (5.6) with respect to the transfer function of a PID controller given in (5.3), we have to obtain the particular values for the parameters Kp, Kd, and Ki of the obtained controller, being:

$$K_p = 2.7$$
$$K_d = 1.2$$
$$K_i = 0.6$$

Chapter 6
System Simulation

In the preceding chapters, we addressed the problem of solving the dynamical equations in the form of differential equations that describe the linear system behavior. However, most of the relationships that define a dynamic system are nonlinear, and moreover, the most of the linear systems treated are a particular case of non-linear systems in limited ranges of operation.

The solution of the nonlinear system cannot be generalized, that is, the conclusions drawn are only valid for the initial conditions and parameters for which they have been determined, and besides cannot in general, be solved analytically, computer simulation usually being the technique employed to get the solution numerically.

A number of useful numerical integration methods of ordinary differential equations will be explained and applied to more realistic and complex nonlinear processes with large numbers of dynamic equations. We will deal only with lumped system, since the treatment of partial differential equations would require more space than is available in this book.

Some examples have been included to help the reader to clarify the methodology used to solve numerically the dynamical equations, which have been also simulated by using both the SIMULINK and SIMSCAPE modeling environments, which make it easier for the reader to understand the development and solution of these examples.

6.1 Simulation Objectives

The simulation technique includes an extensive collection of methods and applications aimed at reproducing the actual behavior of nonlinear systems on a digital computer with appropriate simulation software.

The computer simulation is applied to a dynamic model previously obtained by using the modeling techniques described previously. Starting from it, is derived a computer model in order to run experiments to improve understanding of system behavior under a set of changing conditions.

Continuous-time systems must undergo a process of discretization that transforms the governing differential equations in discrete equations. They use numerical integration methods, whose aim is the replacement of the derivatives of system variables by approximate expressions.

In general, we are going to consider nth-order system represented by a set of n state variable equations, as follows

$$\frac{d\bar{x}}{dt} = \bar{f}\big(\bar{x}(t), \bar{u}(t)\big) = \bar{f}(\bar{x}(t), t) \tag{6.1}$$

where $\bar{x}(t)$ and $\bar{u}(t)$ represent the state vector and the input vector respectively, and $\bar{f}(\bar{x}(t), t)$ is, in general, a non-linear function, and represents the dynamic model that approximates the behavior of the system under study.

The objective of the simulation techniques is obtaining a sequence of values of the state vector $\bar{x}(t_0), \bar{x}(t_1), \ldots, \bar{x}(t_i), \bar{x}(t_{i+1})$ which approximates the solution of the ordinary differential equation (ODE) system in fig. 6.1, by using the parameter defined as integration step is given by $T = t_{i+1} - t_i$.

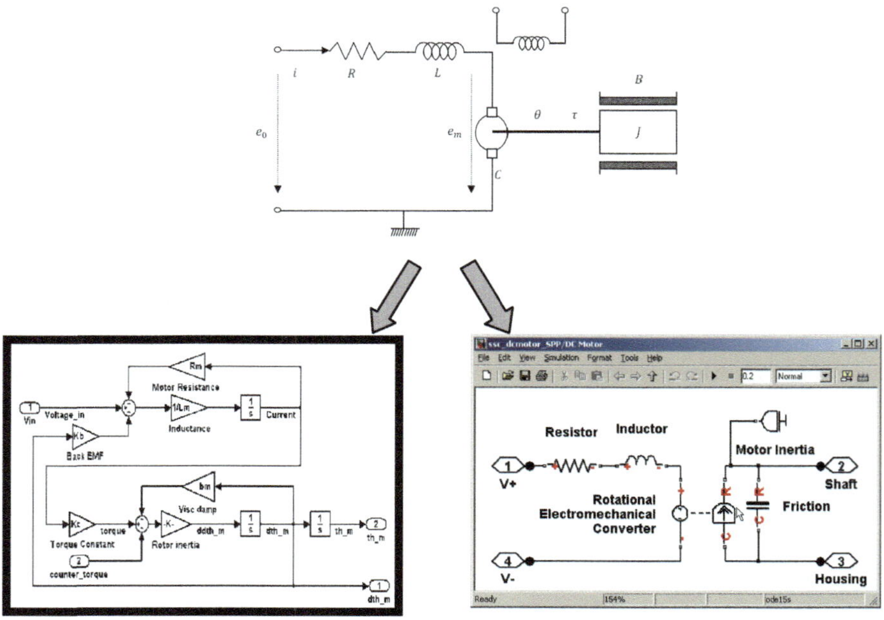

(a) (b)

Fig. 6.1 Computer simulation of a DC motor by using (a) SIMULINK and (b) SIMSCAPE

A computer simulation must embody several components. In first place, the structure of the dynamic model to be simulated must be known, as also the set of model parameters and initial conditions. In second place, the set of input signals must also be embodied, and a set of output signals must be explicitly defined in order to follow the system evolution. Finally, a simulation run time control system must be included in order to select the numerical integration method used and the value of its associated parameters, integration step, error tolerances, etc. We will particularize the computer simulation task by employing simulation software as SIMULINK or SIMSCAPE, using both block diagrams and physical elements interconnected to synthesize the model equations (fig. 6.1).

6.2 Numerical Methods

The solution of ordinary differential equations (ODEs) on a digital computer involves numerical integration. These techniques can be grouped into two broad classes, termed explicit and implicit integration methods.

Common to all these methods, the resolution of the ordinary differential equations which describe the system dynamics of (6.1) by integration between points t_{i-r} and t_{i+1} as

$$\int_{\bar{x}(t_{i-r})}^{\bar{x}(t_{i+1})} d\bar{x} = \int_{t_{i-r}}^{t_{i+1}} \bar{f}(\bar{x}(\tau),\tau)d\tau \tag{6.2}$$

$$\bar{x}(t_{i+1}) = \bar{x}(t_{i-r}) + \int_{t_{i-r}}^{t_{i+1}} \bar{f}(\bar{x}(\tau),\tau)d\tau \tag{6.3}$$

where t_i indicates the actual simulation time, while the r index denotes the number of integration steps used to approximate this integral.

Based on the approximation of $\bar{f}(\bar{x}(t),t)$ in the interval of integration and the number of integration steps selected, arise the different integration methods.

6.2.1 Explicit Numerical Methods

These numerical methods characterized by the state variable $\bar{x}(t_{i+1})$ in (6.3) is a function of the variable values at step i and earlier. These methods are straightforward and easy to program, but present as a major disadvantage that a small step size T must be used so that the solution converge.

Among these methods, the best known is the method of explicit Euler, given by the recursive equation

$$\bar{x}(t_{i+1}) = \bar{x}(t_i) + T\bar{f}(\bar{x}(t_i),t_i), \tag{6.4}$$

which assumes that the function is constant $\bar{f}(\bar{x}(t),t) = \bar{f}(\bar{x}(t_i),t_i)$ for $t_i \leq t \leq t_{i+1}$, and only one integration step ($r = 0$) is used to approximate this integral.

Also belong to this family the Runge–Kutta methods, characterized by evaluating the $\bar{f}$ values at intermediate points inside de integration step. The number of points used in each step of integration determines the order of method. The most common version of the method uses four points in each integration step to approximate the integral in (6.3), and is given by

$$\bar{x}(t_{i+1}) = \bar{x}(t_i) + (\frac{1}{6}\bar{m}_1 + \frac{1}{3}\bar{m}_2 + \frac{1}{3}\bar{m}_3 + \frac{1}{6}\bar{m}_4) \tag{6.5}$$

$$\begin{aligned} \bar{m}_1 &= T\bar{f}(\bar{x}(t_i),t_i) \\ \bar{m}_2 &= T\bar{f}\left(\bar{x}(t_i) + \frac{\bar{m}_1}{2}, t_i + \frac{T}{2}\right) \\ \bar{m}_3 &= T\bar{f}\left(\bar{x}(t_i) + \frac{\bar{m}_2}{2}, t_i + \frac{T}{2}\right) \\ \bar{m}_4 &= T\bar{f}(\bar{x}(t_i) + \bar{m}_3, t_i + T) \end{aligned} \tag{6.6}$$

6.2.2 *Implicit Numerical Methods*

These numerical methods are characterized by the state variable $\bar{x}(t_{i+1})$ in (6.3) is a function of the variable values at step $i + 1$ and earlier. These methods lead to the resolution of non-linear implicit equations for obtaining the $\bar{x}(t_{i+1})$ values, which implies an additional computational effort, although they account for greater stability in the convergence of the solution.

In the same way as we described formerly, the best known is the method of implicit Euler, given by the recursive equation

$$\bar{x}(t_{i+1}) = \bar{x}(t_i) + T\bar{f}(\bar{x}(t_{i+1}),t_{i+1}) \tag{6.7}$$

which assumes that the function is constant $\bar{f}(\bar{x}(t),t) = \bar{f}(\bar{x}(t_{i+1}),t_{i+1})$ for $t_i \leq t \leq t_{i+1}$, and only one integration step ($r = 0$) is used to approximate this integral.

Table 6.1 List of both explicit and implicit Runge-Kutta numerical methods

Integration Method	Algorithm
Runge–Kutta 1 (Explicit Euler)	$\bar{m}_1 = T\bar{f}(\bar{x}(t_i), t_i)$ $\bar{x}(t_{i+1}) = \bar{x}(t_i) + \bar{m}_1$
Rung–Kutta 2 Explicit (Heun)	$\bar{m}_1 = T\bar{f}(\bar{x}(t_i), t_i)$ $\bar{m}_2 = T\bar{f}\left(\bar{x}(t_i) + \frac{\bar{m}_1}{2}, t_i + \frac{T}{2}\right)$ $\bar{x}(t_{i+1}) = \bar{x}(t_i) + \frac{\bar{m}_1}{2} + \frac{\bar{m}_2}{2}$
Rung–Kutta 4 Explicit	$\bar{m}_1 = T\bar{f}(\bar{x}(t_i), t_i)$ $\bar{m}_2 = T\bar{f}\left(\bar{x}(t_i) + \frac{\bar{m}_1}{2}, t_i + \frac{T}{2}\right)$ $\bar{m}_3 = T\bar{f}\left(\bar{x}(t_i) + \frac{\bar{m}_2}{2}, t_i + \frac{T}{2}\right)$ $\bar{m}_4 = T\bar{f}(\bar{x}(t_i) + \bar{m}_3, t_i + T)$ $\bar{x}(t_{i+1}) = \bar{x}(t_i) + (\frac{1}{6}\bar{m}_1 + \frac{1}{3}\bar{m}_2 + \frac{1}{3}\bar{m}_3 + \frac{1}{6}\bar{m}_4)$
Rung–Kutta 1 (Implicit Euler)	$\bar{m}_1 = T\bar{f}(\bar{x}(t_{i+1}), t_{i+1})$ $\bar{x}(t_{i+1}) = \bar{x}(t_i) + \bar{m}_1$
Rung–Kutta 2 Implicit (Trapezoidal)	$\bar{m}_1 = T\bar{f}(\bar{x}(t_i), t_i)$ $\bar{m}_2 = T\bar{f}\left(\bar{x}(t_i) + \frac{\bar{m}_1}{2}, t_i + T\right)$ $\bar{x}(t_{i+1}) = \bar{x}(t_i) + \frac{\bar{m}_1}{2} + \frac{\bar{m}_2}{2}$

In the same way, there exist implicit Runge-Kutta methods characterized by evaluating the $\bar{f}$ values at intermediate points inside the integration step, including the $f(\bar{x}(t_{i+1}), t_{i+1})$ value.

In table 1 are listed with both the explicit and the implicit Runge-Kutta methods commonly used for different orders (RK-n) depending on the number of points to be used to approximate the integral.

Example 6.1

Derive the approximate solution corresponding to the dynamical equation given by

$$\frac{dx}{dt} = -\mu x$$

by applying both the explicit and implicit Euler method. Simulate the system by using both methods in SIMULINK for varying integration steps of $T = 0.5$, $T = 1$ and $T = 2$ for $\mu = 1$, also comparing them with the exact solution $x(t) = x(t_0)e^{\mu t}$.

Solution

Applying (6.4) for $f(x) = -\mu x$ we get

$$x(t_{i+1}) = x(t_i) - \mu\, T\, x(t_i)$$

$$x(t_{i+1}) = (1 - \mu\, T)x(t_i)$$

Writing this equation in terms of the initial condition $x(t_0)$, and solving iteratively we obtain the solution points for the explicit solution

$$x(t_{i+1}) = (1 - \mu\, T)^{i+1}x(t_0)$$

which will be stable only if $\mu T < 1$.

Following the same approach, applying (6.7) $f(x) = -\mu x$ we have

$$x(t_{i+1}) = x(t_i) - \mu\, T\, x(t_{i+1})$$

$$x(t_{i+1}) = \frac{1}{1 + \mu\, T}x(t_i)$$

Writing this equation in terms of the initial condition $x(t_0)$, and solving iteratively we obtain the solution points for the implicit solution

$$x(t_{i+1}) = (\frac{1}{1 + \mu\, T})^{i+1}x(t_0)$$

which will be stable no matter the value for T be selected.

In fig. 6.2 it is shown the simulation diagram corresponding to the dynamis system defined in this example. The output responses for x(0) = 4 are shown in fig. 6.3 for increasing values of T = 0.5, T = 1 and T = 2 for the system numerically integrated using explicit Euler ('ode1' in SIMULINK). It can be shown how the approximated solution diverges from the exact solution as long as $\mu T = T < 1$ (fig. 6.4).

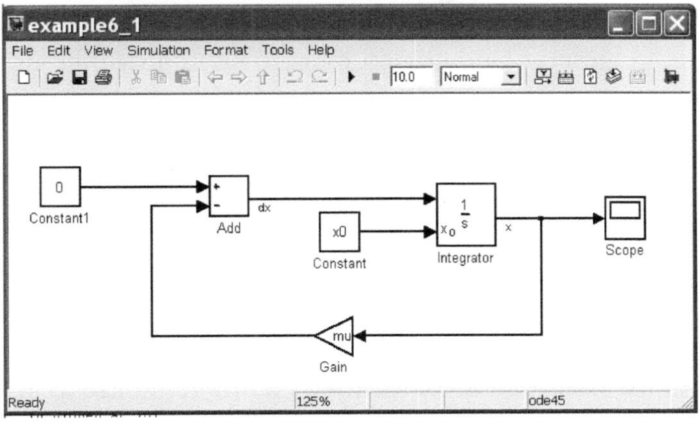

Fig. 6.2 SIMULINK diagram corresponding to example 6.1.

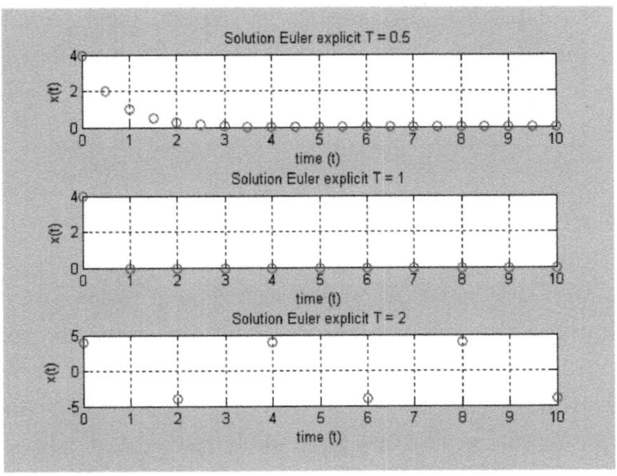

Fig. 6.3 Output response obtained by using the explicit Euler method for varying T.

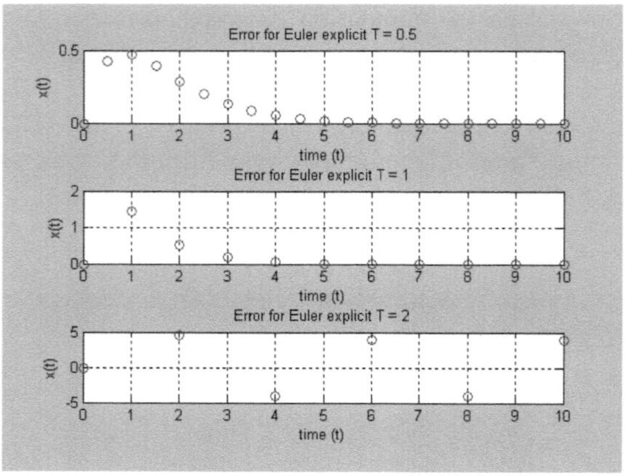

Fig. 6.4 Output error obtained by using the explicit Euler method for varying T.

6.2.3 Multistep Numerical Methods

Multistep methods are characterized by the state variable $\bar{x}(t_{i+1})$ in (6.3) is a function of the variable values $\bar{x}(t_i), \bar{x}(t_{i-1}), \dots, \bar{x}(t_{i-r})$. These methods can be grouped in a similar way as we did before, in explicit and implicit integration methods, accounting for the Adams-Bashforth (AB) and Adams-Moulton (AM) methods. Normally these methods need the solution at several preceding time points to compute the current solution ('starting problem').

In table 6.2 are listed both the explicit AB-n and the implicit AM-n methods commonly used for different order *n*, depending on the number of steps to be used to approximate the integral (for example, 'ode113' in SIMULINK with up to 13 steps).

6.2.4 Integration Step Selection

The integration step size must be properly chosen in order to avoid unstabiliy in the ODE's solution. In nearly all the cases, there is usually some information regarding the system time constants. Then, the integration step must be chosen as ¼ of the smallest time constant of the system, in case the integration step is fixed.

However, the integration step can be made variable, depending on the estimated integration error as the difference in the evaluation of xi +1 with two different methods in each step of integration, for example RK-4 and RK-5 ('ode45' in SIMULINK) Thus, if the error is less than a tolerance, the step of integration is increased, and vice versa.

Table 6.2 List of both Adams-Bashforth and Adams-Moulton numerical methods

Integration Method	Algorithm
Adams-Bashforth 1 (Explicit Euler)	$\bar{m}_1 = T\bar{f}(\bar{x}(t_i), t_i)$ $\bar{x}(t_{i+1}) = \bar{x}(t_i) + \bar{m}_1$
Adams-Bashforth 2	$\bar{m}_1 = T\bar{f}(\bar{x}(t_i), t_i)$ $\bar{m}_2 = T\bar{f}(\bar{x}(t_{i-1}), t_{i-1})$ $\bar{x}(t_{i+1}) = \bar{x}(t_i) + \frac{3}{2}\bar{m}_1 - \frac{\bar{m}_2}{2}$
Adams-Bashforth 3	$\bar{m}_1 = T\bar{f}(\bar{x}(t_i), t_i)$ $\bar{m}_2 = T\bar{f}(\bar{x}(t_{i-1}), t_{i-1})$ $\bar{m}_3 = T\bar{f}(\bar{x}(t_{i-2}), t_{i-2})$ $\bar{x}(t_{i+1}) = \bar{x}(t_i) + \frac{23}{12}\bar{m}_1 - \frac{4}{3}\bar{m}_2 + \frac{5}{12}\bar{m}_2$
Adams-Moulton 1 (Implicit Euler)	$\bar{m}_1 = T\bar{f}(\bar{x}(t_{i+1}), t_{i+1})$ $\bar{x}(t_{i+1}) = \bar{x}(t_i) + \bar{m}_1$
Adams-Moulton 2	$\bar{m}_1 = T\bar{f}(\bar{x}(t_{i+1}), t_{i+1})$ $\bar{m}_2 = T\bar{f}(\bar{x}(t_i), t_i)$ $\bar{x}(t_{i+1}) = \bar{x}(t_i) + \frac{\bar{m}_1}{2} + \frac{\bar{m}_2}{2}$
Adams-Moulton 3	$\bar{m}_1 = T\bar{f}(\bar{x}(t_{i+1}), t_{i+1})$ $\bar{m}_2 = T\bar{f}(\bar{x}(t_i), t_i)$ $\bar{m}_3 = T\bar{f}(\bar{x}(t_{i-1}), t_{i-1})$ $\bar{x}(t_{i+1}) = \bar{x}(t_i) + \frac{5}{12}\bar{m}_1 + \frac{2}{3}\bar{m}_2 - \frac{1}{12}\bar{m}_2$

In spite of this, some systems exhibit time constants very different from one another, so that the integration methods or even the variable step size algorithm described will not work. These systems are called stiff systems.

A large number of implicit methods are well suited for stiff systems, in particular the Gear method ('ode15s' in SIMULINK) and Rosenbrock ('ode23s' in SIMULINK).

Table 6.3 Characteristics of the numerical methods handled in SIMULINK

Integration Method	Characteristics
Explicit RK-*n* (*ode1- ode5*)	Explicit Runge-Kutta one-step solver, with fixed integration step. Useful when time constant is known. More accuracy for higher *n* values
Explicit RK4-RK5 (*ode45*)	Explicit Runge-Kutta one-step solver, with variable integration step. Is the default solver, and the first method to try for most systems.
Explicit RK2-RK3 (*ode23*)	Explicit Runge-Kutta one-step solver, with variable integration step. Is more efficient than *ode45* for coarse tolerances
AB-AM (*ode113*)	Adams-Bahforth Adams-Moulton multistep solver, with variable integration step. Is more efficient than *ode45* for stringent tolerances, useful when high accuracy is demanded
Gear Method (*ode15s*)	Gear based variable order and multistep solver. Is efficient when some kind of stiffness is present
Rosenbrock Formula (*ode23s*)	Rosenborck order 2 formula one-step solver. Is efficient when some kind of stiffness is present and is more efficient than *ode23s* for coarse tolerances

The SIMULINK environment incorporates most of the integration methods presented, both explicit (Runge–Kutta and Adams–Bashfort) and implicit (Gear and Rosenbrock) including one-step and multistep ones, and table 6.3 shows the main characteristics of the referred methods.

6.3 Systems Simulation in SIMULINK

SIMULINK is a graphical, block oriented simulation language for modeling and simulation of dynamics systems. This computer package is an extension to MATLAB and can also interface with in order to provide it with flexibility.

In SIMULINK, systems are drawn on screen as block diagrams by selecting the basic components from a large library of component blocks, among which we include continuous, discontinuous, math operators, signal routing, inputs, and outputs blocks.

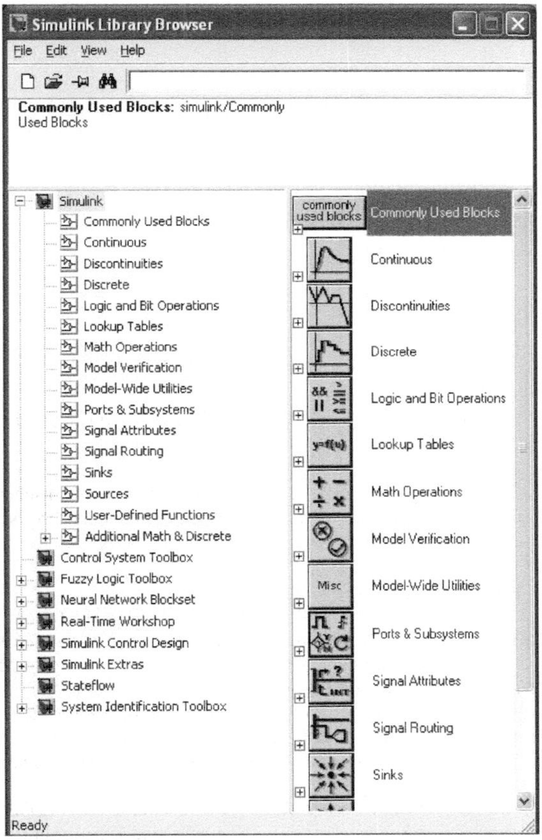

Fig. 6.5 The Library browser of the SIMULINK program.

In SIMULINK, a model is a collection of blocks which, in general, represents a dynamic system. Blocks are used to generate, modify, combine, output, and display signals, and can be modified by double-clicking on them. Lines are used to transfer signals from one block to another. These blocks are arranged in Block

Libraries which are accessed in the main SIMULINK window (fig. 6.5). Block parameters can be defined from the MATLAB workspace.

The process of numerically integrating of the dynamics equations is essential for the simulation task, and hence the integrator blocks form the core of the simulation diagram.

The construction of the block diagram starts with the assembly of lines of integrators for each of the state variables, gathering the adequate elements from the Block Libraries to build its derivative functions associated. The initial conditions for the state variables will be set within the integrators, and the outputs of the integrators are the state variables themselves.

In this way, for a nonlinear dynamic system as defined by (6.1) in general, by integrating both sides we obtain

$$\bar{x}(t) = \int \bar{f}\big(\bar{x}(\tau), \bar{u}(\tau)\big)\, d\tau \tag{6.8}$$

expression which represents the lines of integrators for $\bar{x}(t)$, whose particular form depends on $\bar{f}\big(\bar{x}(\tau), \bar{u}(\tau)\big)$ specifically.

Once the simulation is performed, the output signals (states) can be displayed graphically, and simulation run-time control window enables the selection of the integration method along with the solver parameters (initial time, final time, step size, error tolerance,...).

Example 6.2

Derive the block diagram for the dynamic system represented by

$$\frac{d^2x}{dt^2} = Ku(t) - B\sqrt{x(t)} + sin(\frac{dx}{dt})$$

with initial conditions $x(0) = x_0$ and $\frac{dx}{dt}(0) = \dot{x}_0$. Run the simulation for $K = 1.5$, $B = 2$, $x_0 = 1$ and $\dot{x}_0 = 0$, for $0 < t < 10$ by using the '*ode45*' solver, and display the output response $x(t)$ to a square input $u(t)$ with unitary amplitude and frequency of 1 Hz.

Solution

Transforming this equation into state space, by defining $x_1 = x$ and $x_2 = \frac{dx}{dt}$ we have

$$\frac{dx_1}{dt} = x_2$$

$$\frac{dx_2}{dt} = Ku(t) - B\sqrt{x_1} + sin(x_2)$$

These two equations define the line of integrators essential to build the SIMULINK model, which form the core of the block diagram. We show in fig. 6.6 the complete simulation block diagram in SIMULINK for this system and fig. 6.7 displays the simulation run-time control window, showing information about initial and final time, integration algorithm, minimum and maximum step size, and error tolerance.

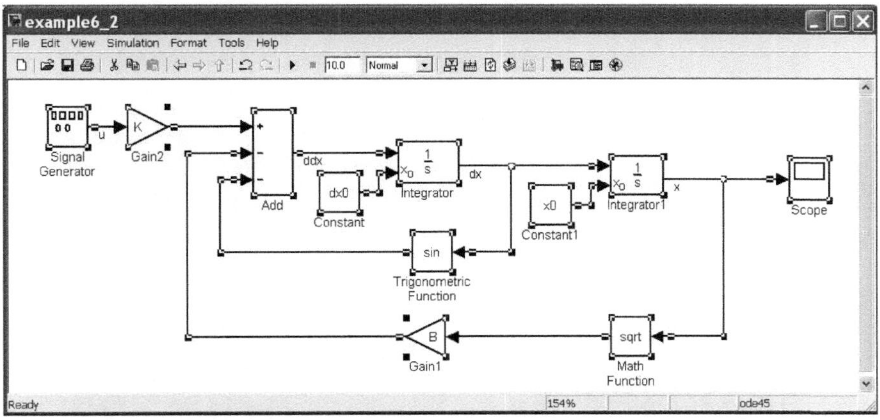

Fig. 6.6 SIMULINK diagram corresponding to example 6.2.

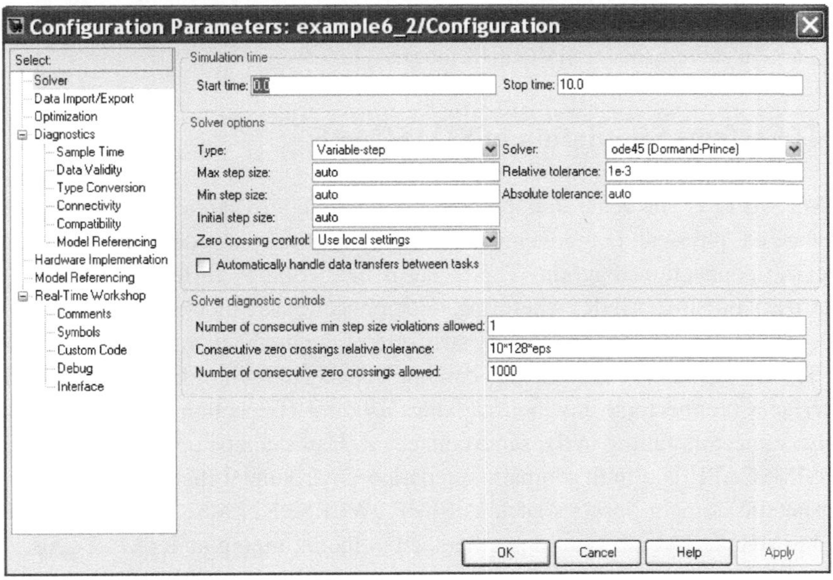

Fig. 6.7 Run-time control settings corresponding to example 6.2.

The parameters K and B together with the initial conditions have been set within the MATLAB workspace

```
>> x0=1;dx0=0;
>> K=1.5;B=2;
```

In fig. 6.8 it is depicted the system response to a square input signal.

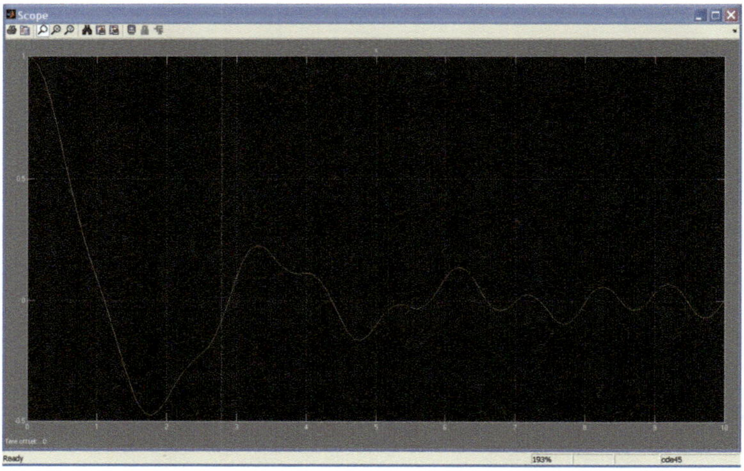

Fig. 6.8 Output response in SIMULINK for example 6.2.

6.4 Systems Simulation in SIMSCAPE

SIMSCAPE is an acausal approach for modeling dynamic systems by using connected physical components in which equations are defined, following a specific connection diagram. This acausal methodology differs from the causal one used in SIMULINK, where the interconnection of blocks reflects rather the calculation procedure than the very structure of the modeled system.

These physical components can be connected through exactly defined interfaces (connectors) in which variables used for connection are defined, so that only connectors falling in the same connector classes can be connected.

SIMSCAPE is a multi-domain simulation environment that can also be linked to specific domain libraries such as SIMPOWERSYSTEMS, SIMMECHANICS, and SIMHYDRAULICS among others, all of them running under MATLAB.

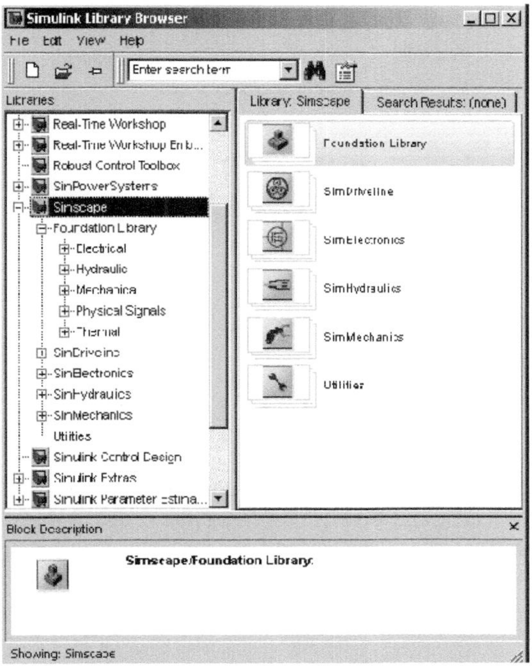

Fig. 6.9 The Physical Library browser of the SIMSCAPE program.

In SIMSCAPE, a model is a collection of physical blocks which must be appropriately connected to define the dynamic system. Physical blocks can be of very distinct nature, ranging from electric, mechanical, hydraulic and thermal, and its constitutive parameters can be modified easily by double-clicking on them. Lines are used to transfer physical signals from one block to another, in contrast with SIMULINK. These physical blocks are arranged in Physical Libraries which are accessed in the main SIMSCAPE window (fig. 6.9), and physical block parameters can be defined from also the MATLAB workspace.

The construction of the physical block diagram starts with the assembly of each one of the physical components which define the system model, gathering the adequate components from the Physical Block Libraries.

The Physical Network approach supports two types of variables (fig. 6.10), effort variables ($\mathrm{e_A}, \mathrm{e_B}$) that are measured with a sensor connected in parallel to a component with respect to a reference, and flow variables (f_{AB}) that are measured with a sensor connected in series to a component.

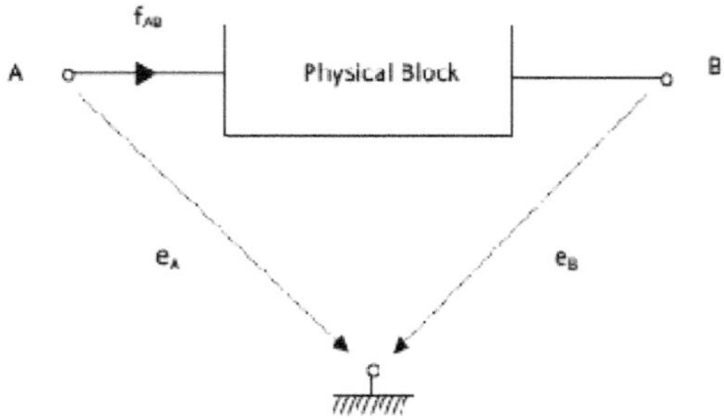

Fig. 6.10 The effort and flow variables for each SIMSCAPE component

There are different ports associated to the SIMSCAPE blocks, namely, conserving ports (■) and signal ports (▻), different form the SIMULINK standard ports (>).

The conserving ports represent physical connections and relate physical variables among physical blocks, and connection lines that connect these ports together are bidirectional lines that carry effort, and flow variables rather than signals. The sum of effort variables in a closed loop is null, while the sum of flow variables flowing into a branch point, equals the sum of all its values flowing out (fig. 6.11).

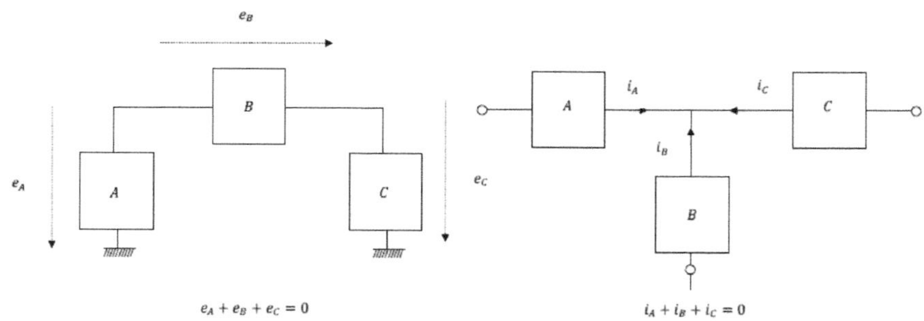

Fig. 6.11 Conservation laws for physical block diagrams in SIMSCAPE

There are several types of conserving ports in SIMSCAPE each one with effort and flow associated variables, which are listed in table 6.4.

The signal ports transmit signals between blocks with units associated with them, usually specified along with the parameter values in the block, unlike SIMULINK signals, which are dimensionless. Associated with these types of ports are the sensor and actuator blocks. SIMSCAPE is equipped with a Physical Signal Library which enables to implement linear and nonlinear operators, functions, lookup tables, etc. In case a relationship is not available in SIMSCAPE to implement a determined component.

In order to manage the signal inputs and outputs to and from the physical block diagram, it is necessary to use converter blocks to connect SIMSCAPE diagrams to SIMULINK sources and scopes. The SIMULINK-PS Converter block connect

Table 6.4 List of physical conserving ports in SIMSCAPE

Physical Port Type	Effort Variable	Flow Variable
Electrical	Voltage	Current
Mechanical Translational	Velocity	Force
Mechanical Rotational	Angular Velocity	Torque
Hydraulic	Pressure	Flow Rate
Thermal	Temperature	Heat Flow

SIMULINK sources or other blocks to the inputs of a SIMSCAPE diagram, and conversely, the PS-SIMULINK Converter block connect outputs of a SIMSCAPE diagram to SIMULINK scopes or other blocks.

Finally, it must be emphasized that 'ode15s' is one of the numerical integration methods allowed by SIMSCAPE to run the model, so that a "Solver Configuration block" must be connected to the diagram by creating a branching point at any physical connection.

Example 6.3

Derive the SIMSCAPE physical diagram for the electric circuit shown in (fig. 6.12), and plot the time response $v_0(t)$ to input signal $i_S(t) = 0.5u_s(t)$ for values of R1 = 0.2, R2 = 1, C = 0.01 and L = 0.1, starting from initial conditions $v_0(0) = 0$ and $\frac{dv_0}{dt}(0) = 0$.

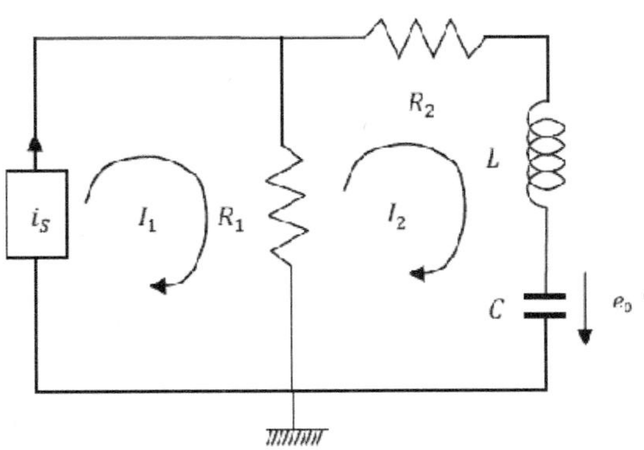

Fig. 6.12 Diagram of components for the electric circuit of example 6.3.

Solution

In first place we gather the electric components necessary to build the circuit, including both the current source, voltage meter and the solver configuration block (see fig. 6.13)

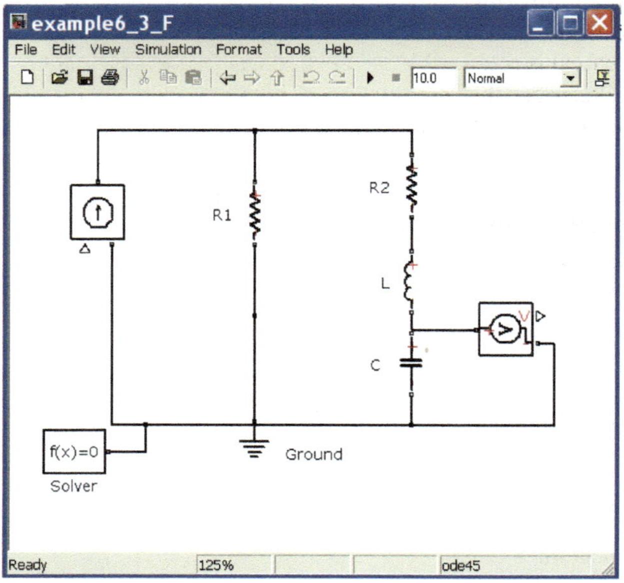

Fig. 6.13 SIMSCAPE diagram corresponding to example 6.3

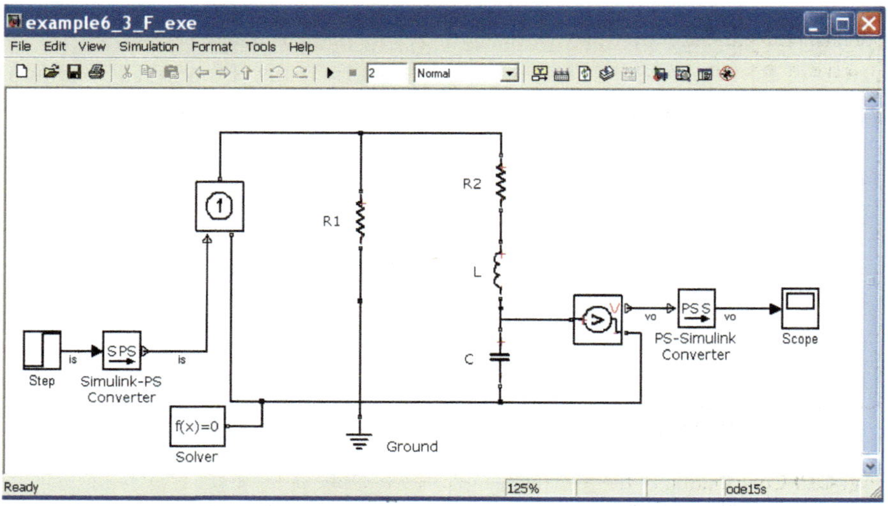

Fig. 6.14 SIMSCAPE diagram corresponding to example 6.3 including input-output signals.

Adding the step input for the source current $i_a(t)$ through the converter SPS and measuring capacitor voltage $v_0(t)$ through a voltmeter to a scope by using the PSS converter, is completed in the SIMSCAPE block diagram (fig. 6.14).

In fig. 6.15 the step response for the particular value of parameters

>> R1=0.2; R2=1; C=0.01; L=0.1 is shown;

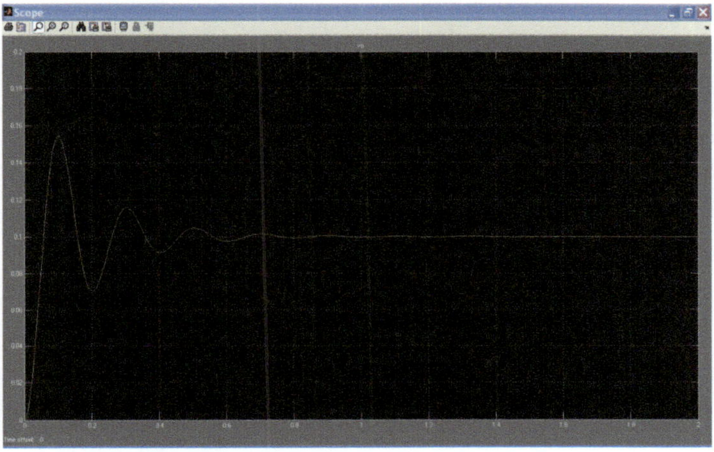

Fig. 6.15 Output response for the SIMSCAPE model of example 6.3

6.5 Applications

In this section we are going to describe more comprehensive applications of SIMULINK and SIMSCAPE simulation environments.

Example 6.4

The system of fig. 6.16 represents a height controller system for two tanks. The dynamic equations of the system include the nonlinear effects of the inlet flow $q_e(t)$ and the unidirectional flow between tanks $q_{12}(t)$

The system equations are given as follows:

$$e(t) = K_t(u(t) - h_2(t))$$

$$q_e(t) = \begin{cases} min\,(K_g e(t), q_{emax}) & e(t) \geq 0 \\ 0 & e(t) < 0 \end{cases}$$

$$q_{12}(t) = \begin{cases} \dfrac{h_1(t) - h_2(t)}{R_{12}} & h_1(t) \geq h_2(t) \\ 0 & h_1(t) < h_2(t) \end{cases}$$

$$q_2(t) = \frac{h_2(t)}{R_2}$$

$$A_1 \cdot \frac{dh_1}{dt} = q_e(t) - q_{12}(t)$$

$$A_2 \cdot \frac{dh_2}{dt} = q_{12}(t) - q_2(t)$$

being h_1 , h_2 ,q_e , q_{12} , q_2 the tank-1 height , tank-2 height, input flow to tank 1, flow from tank-1 to tank-2, and output flow from tank-2, and R_{12}, R_2, A_1, A_2 the hydraulic resistances and capacitances, together K_t, K_g amplifier gains and q_{1max}a limiting value for q_e.

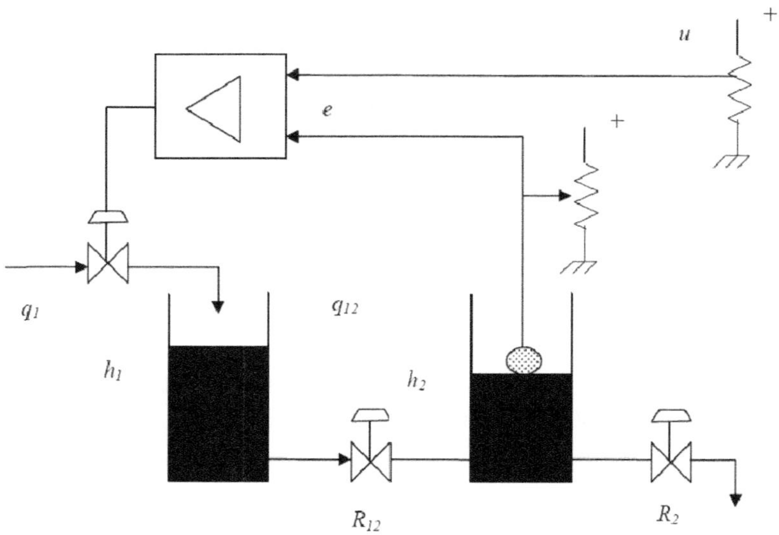

Fig. 6.16 Height control system for a two tank system.

Simulate the dynamic behavior in SIMULINK for a reference input pulse of height 2 m and period 20 s (50% OFF 50% ON) for values of $K_t = 1$, $K_g = 5$, $R_{12} = 4$, $R_2 = 4$, $A_1 = 0.8$, $A_2 = 0.2$, $q_{1max} = 5$, plotting the evolution of $h_1(t)$ $h_2(t)$, $q_e(t)$,$q_{12}(t)$, by designing a block "subsystem" for each tank.

Solution

In first place we build the line of integrators for this system (fig. 6.17)

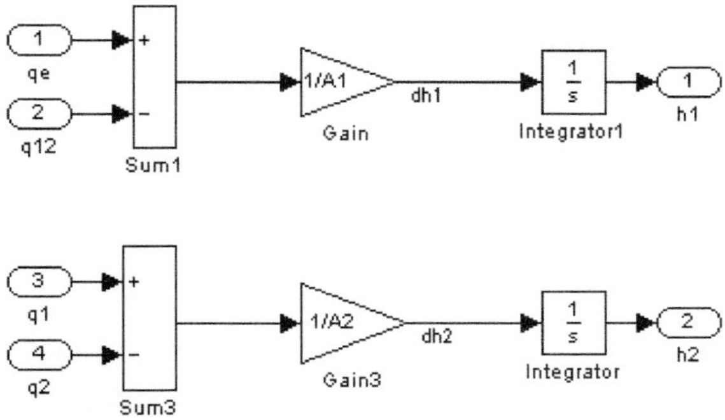

Fig. 6.17 Lines of integrators for the two tank system (encoded as 'subsytems')

Then, we implement the state equations by using the appropriate blocks from the Block Libraries, embedding the loading tank dynamics into 'subsystems' as indicated above, resulting in the block diagram of (fig. 6.18).

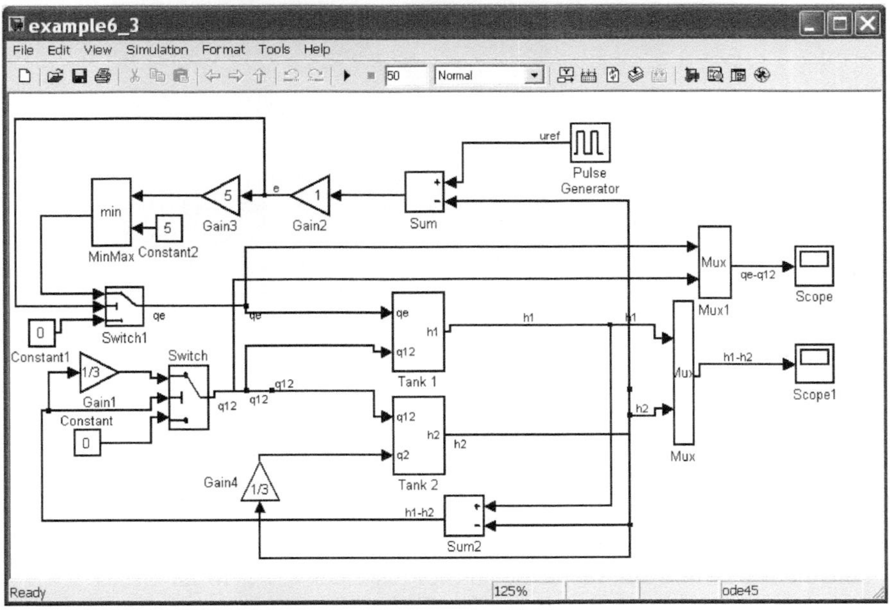

Fig. 6.18 SIMULINK diagram corresponding to example 6.4.

The system achieves the control objective at tank-2 when a pulse height input is applied (fig. 6.19), while maintaining the flowrate limitation conditions (fig. 6.20).

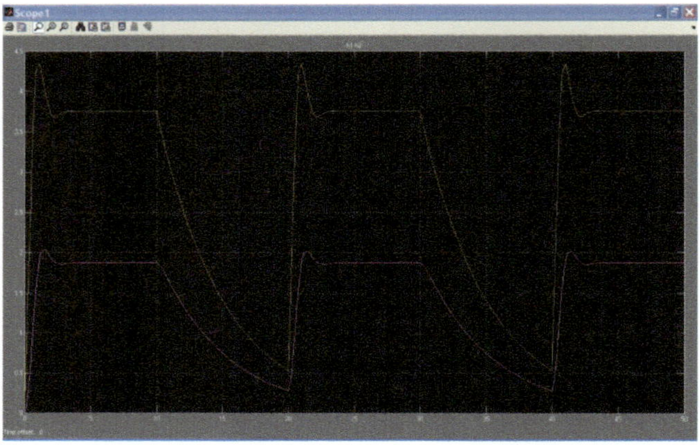

Fig. 6.19 Heights h_1 and h_2 for a pulse change for example 6.4

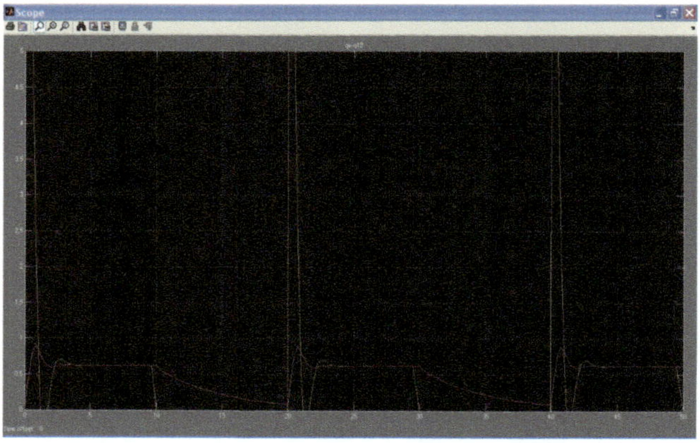

Fig. 6.20 Flowrates q_e and q_{12} for a pulse change for example 6.4.

Example 6.5

The diagram of fig. 6.21 shows a position servomechanism used for the car guidance. The aim is to position the mechanical load on c(t) (wheel angle) according to the reference position r(t) (steering wheel angle). For this, two potentiometers measure the error position, and convert it to an electrical signal proportional $e_v(t)$ through the constant K_p.

A controller based on O.A. amplifies this signal and feeds a DC motor with resistanceR_a and inductance L_a . The motor produces a torque $T(t)$ proportional to the current $i_a(t)$ through the gain K_t, and a rotation speed of $\omega(t)$ proportional through K_t to the counter-electromotive voltage $e_b(t)$. The torque has to overcome the inertia J of the mechanical load and of the engine itself plus the friction through the gear with ratio N.

The electrical diagram of the O.A. is enclosed in (fig. 6.22), being constituted by the four resistances R_1, R_2, R_3, R_4, with an inversion stage included

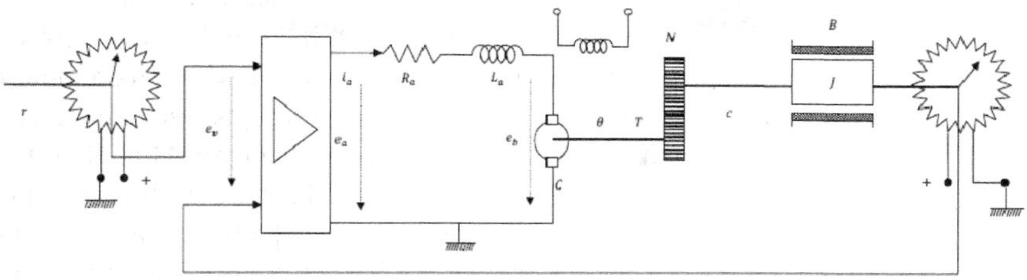

Fig. 6.21 Diagram of the servomechanism of example 6.5.

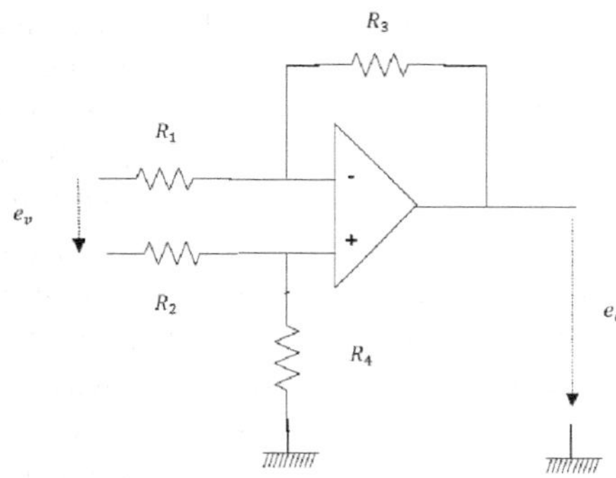

Fig. 6.22 Diagram of the controller of example 6.5.

Simulate the servomechanism in SIMSCAPE for parameter values of $K_p = \frac{12}{\pi}$, $K_t = 0.06$, $R_a = 0.2$, $L_a = 0.01$, $b = 0.0004$, $J = 0.005$, $N = 100$, $R_1 = R_2 = 1$, $R_3 = R_4 = 10$, by using "subsystems" blocks, specifically, controller, motor, and sensor subsystems, obtaining the output response $c(t)$ for a steering step command $r(t)$.

Solution

In first place we have to fill up each subsystem block with the constitutive blocks, starting from an empty block, with a proper number of input and output ports. These ports have to be added afterwards, by dragging them from the "Utilities Library" of SIMSCAPE and eliminating at the same time the SIMULINK port (fig. 6.23).

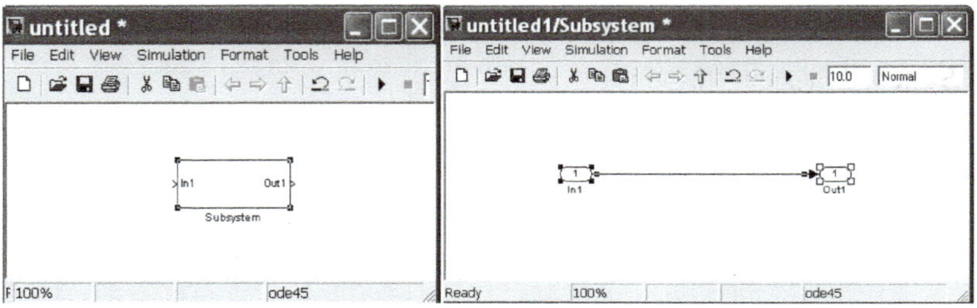

Fig. 6.23 Detail of a subsystem block in SIMULINK

In order to model the Motor Subsystem, we need two input ports and two output ports according to the servomechanism configuration (fig. 6.24).

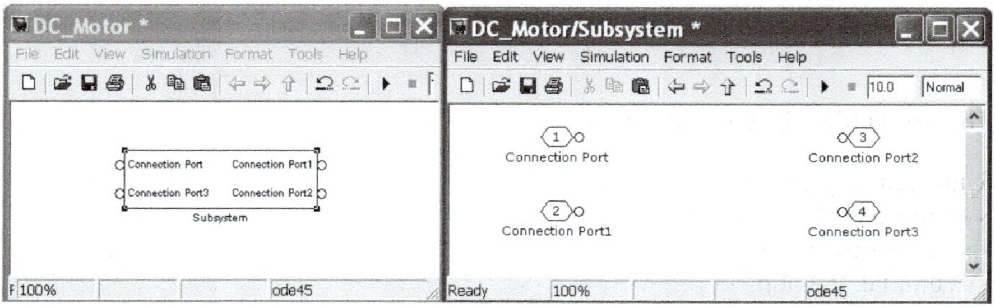

Fig. 6.24 Transformation of a Subsystem block in SIMSCAPE

Then, we have to add the necessary blocks which form part of the DC motor, according to the servomechanism layout. For this task, we will use the Electric and Mechanical Libraries of SIMSCAPE where these blocks are available (fig. 6.25 and fig. 6.26).

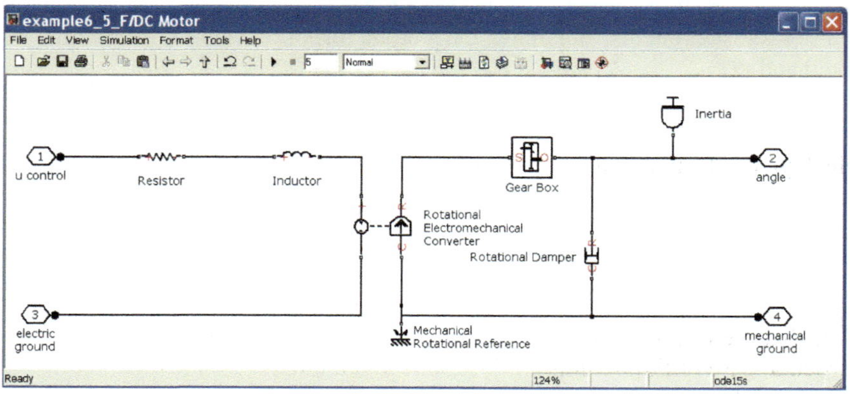

Fig. 6.25 Elements constitutive of the Motor Subsystem in SIMSCAPE

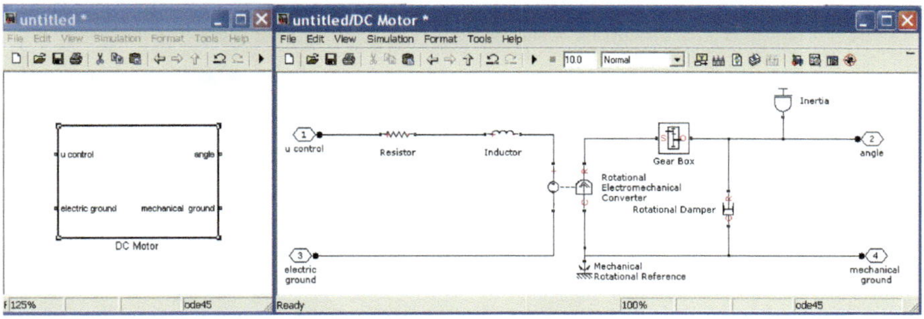

Fig. 6.26 Modeling the Motor Subsystem in SIMSCAPE.

In a similar way, we proceed to the modeling of the controller subsystem by selecting the adequate electric components from the Electric Library (fig. 6.27).

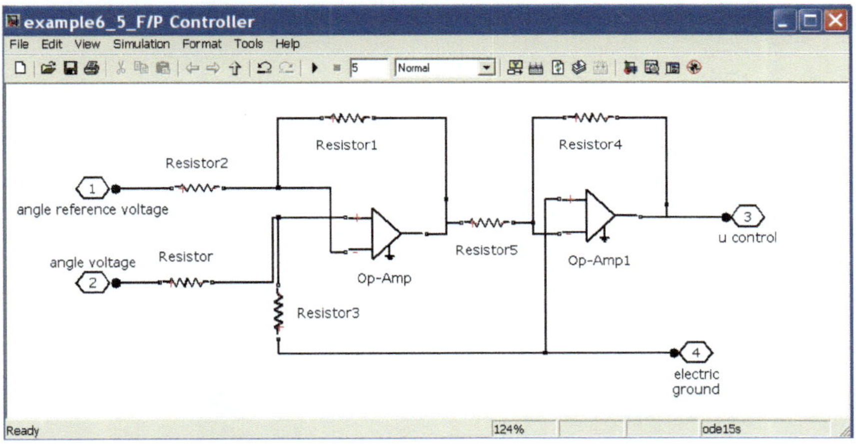

Fig. 6.27 Modeling the Controller Subsystem in SIMSCAPE.

In addition to this, it is also necessary to realize the modeling, both of the angle sensor and of the inverse angle sensor, in order to convert the wheel angle to voltage (angle sensor), and the steering wheel angle to reference voltage (virtual angle sensor) (fig. 6.28).

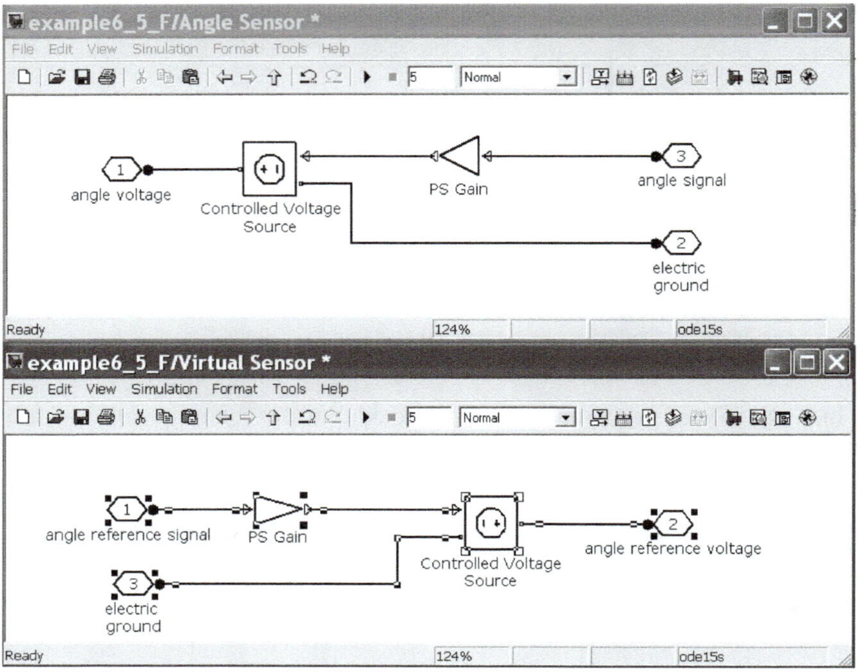

Fig. 6.28 Modeling the Angle Sensor and Virtual Angle Sensor Subsystems in SIMSCAPE.

In fig. 6.29 the complete block diagram obtained linking the subsystems previously modeled following the servomechanism design is shown.

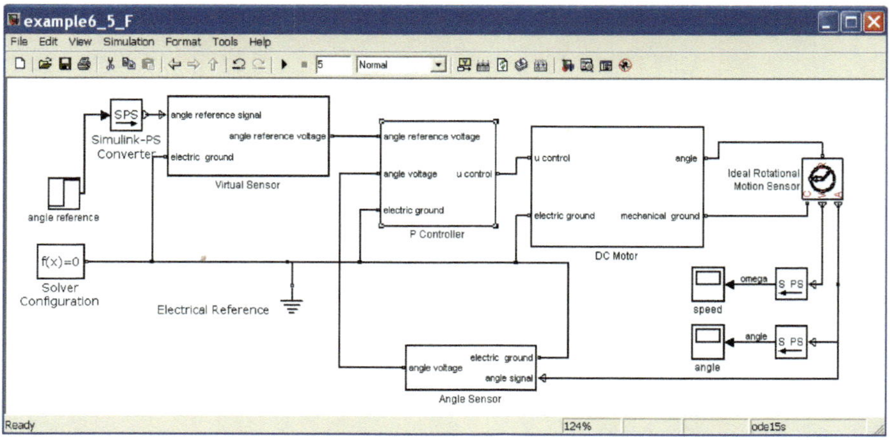

Fig. 6.29 Modeling the complete servomechanism system in SIMSCAPE.

The time response for a unit step reference change in the steering wheel is depicted in fig. 6.30, showing the adequate tracking for the wheel angle. The step response is also included in fig. 6.31.

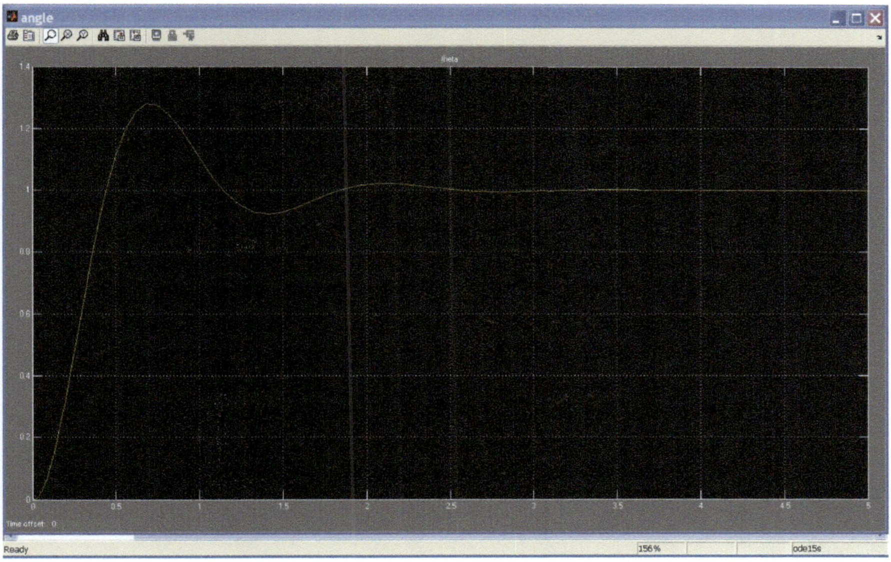

Fig. 6.30 Wheel angle $c(t)$ for a pulse change $r(t)$ for example 6.5.

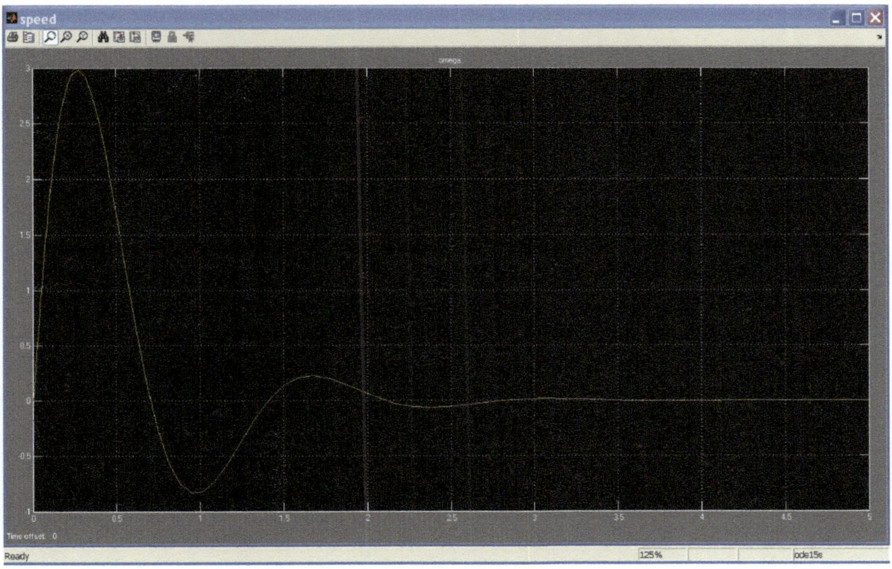

Fig. 6.31 Wheel speed $\dot{c}(t)$ for a pulse change $r(t)$ for example 6.5.

Appendix A: The MATLAB System Control Toolbox

A.1 Creation of Models

The System Control Toolbox of system MATLAB includes commands for the creation of four basic types of models for linear time invariant (LTI) systems (fig. A.1):

- Models Transfer Function (TF)
- Models of zeros, poles, and gain (ZPK)
- State space models (SS)

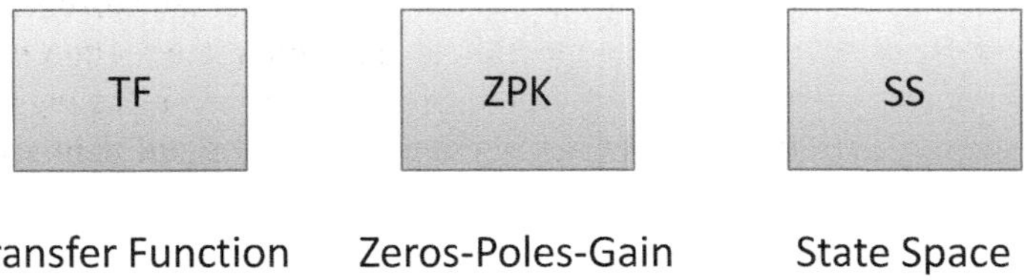

Fig. A.1 Creation of models for linear time invariant (LTI) systems.

These functions take model data as input and return objects that include this data in single MATLAB variables.

A.2 Transfer *Function*

A model described as a transfer function (TF) is defined by their polynomial of the numerator and denominator, that is,

$$H(s) = \frac{p_1 s^n + p_2 s^{n-1} + \cdots + p_{n+1}}{q_1 s^m + q_2 s^{m-1} + \cdots + q_{m+1}}$$

where $p_1, p_2 \cdots p_{n+1}$ are the coefficients of the numerator, while $q_1, q_2 \cdots, q_{m+1}$ are the coefficients of the denominator. Both polynomials are specified by a vector of coefficients $[p_1, p_2 \cdots p_{n+1}]$ and $[q_1, q_2 \cdots, q_{m+1}]$.

Therefore, the model can be represented by the transfer function of a SISO system (Single Input Single Output) by specifying the polynomials for the numerator and denominator as inputs to the command *tf*.

Example

$$H(s) = \frac{s}{s^2 + 2s + 10}$$

```
>> num = [1 0];        % Numerador -> s
>> den = [1 2 10];    %Denominador:-> s^2 + 2s + 10
>> H = tf (num,den)

Transfer function:
      s
--------------
s^2 + 2 s + 10
```

Alternatively this model can be specified as an expression involving the 's' term by defining

```
>> s = tf('s');     % Create Laplace variable
>> H = s/(s^2+2*s+10)

Transfer function:
      s
--------------
s^2 + 2 s + 10
```

In both cases, an object H representing the system model is created in MATLAB.

```
>> H

Transfer function:
      s
--------------
s^2 + 2 s + 10
```

A.3 Zeros, Poles, and Gain

A model can also be described in the form of zeros, poles and gain (ZPK), which is a factored form to express a transfer function model, that is

$$H(s) = k\frac{(s - z_1) \cdots (s - z_n)}{(s - p_1) \cdots (s - p_m)}$$

where K is the gain, $z_1 \cdots z_n$ are the zeros of $H(s)$ and $p_1 \cdots p_m$ are the poles of $H(s)$.

In this format, a model is characterized by its gain value (K), its zeros (which are the roots of the numerator polynomial) and its poles (which are the roots of the denominator polynomial) and is represented by using the command *zpk*.

Example

$$H(s) = 2\frac{(s+1)}{(s-2)(s^2-2s+2)}$$

```
>> z = [-1]; % Zeros
>> p = [2 1+i 1-i]; % Polos
>> k = 2; % Gain
>> H = zpk(z,p,k)

Zero/pole/gain:
       2 (s+1)
---------------------
(s-2) (s^2  - 2s + 2)
```

Alternatively this model can be specified as an expression involving the 's' term by defining

```
>> s = zpk('s');
>> H = 2*(s+1)/((s-2)*(s^2-2*s+2))

Zero/pole/gain:
       2 (s+1)
---------------------
(s-2) (s^2  - 2s + 2)
```

In both cases, an object H representing the system model is created in MATLAB.

```
>> H

Zero/pole/gain:
       2 (s+1)
---------------------
(s-2) (s^2  - 2s + 2)
```

A.4 State Space

A model can also be described in the form of state space (SS) starting from the linear differential equations that describe the dynamics of the system. We use the matrices A, B, C, and D of the state space to characterize this model, in the form

$$\frac{dx}{dt} = A\,x + B\,u$$
$$y = C\,x + D\,u$$

where x is the state vector, u and y are the input and output vectors.

In this format, a model is characterized by the four matrices A, B, C and D, being represented by using the command *ss*, which derives a system object to represent it.

Example

$$\frac{dx}{dt} = Ax + BI \quad ; \quad A = \begin{bmatrix} 0 & 1 \\ -5 & -2 \end{bmatrix} \quad B = \begin{bmatrix} 0 \\ 3 \end{bmatrix}$$

$$y = Cx + Du \quad ; \quad C = [1 \quad 0] \quad D = [0]$$

```
>> A = [0 1;-5 -2];
>> B = [0; 3];
>> C = [1 0];
>> D = 0;
>> H = ss(A,B,C,D)
a =

        x1  x2
   x1    0   1
   x2   -5  -2


b =
        u1
   x1    0
   x2    3

c =
        x1  x2
   y1    1   0

d =
        u1
   y1    0

Continuous-time model.
```

A.5 Access to Model Data

The recovering of data from a system object is made by applying the *tfdata*, *zpkdata* and *ssdata* commands (fig. A.2), using the format

```
>> [num,den] = tfdata (sistema, 'v')
>> [z,p,k] = zpkdata(sistema,'v')
>> [A,B,C,D] = ssdata(sistema)
```

where the argument 'v' enables the return of data vectors instead of arrays of cells.

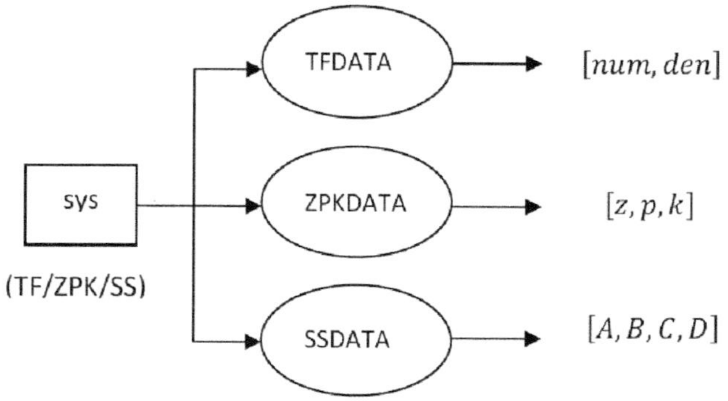

Fig. A.2 Creation of models for linear time invariant (LTI) systems.

Example

Given the SISO transfer function in MATLAB by

```
>> h = tf([1 1],[1 2 5])
```

extract the coefficients of the numerator and denominator.

We would have to apply

```
>> [num,den] = tfdata(h,'v')

    num =
         0    1    1

    den =
         1    2    5
```

A.6 Building Complex Models

There are a number of functions to help build complex models (fig. A.3), among which are included:

- Serial and parallel connections (*series, parallel*)
- Feedback connections (*feedback*)
- Concatenation ([,], [;] and *append*)

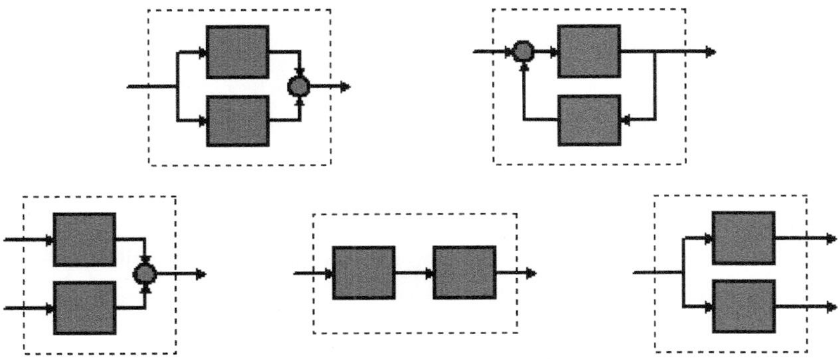

Fig. A.3 Different connection configurations between models

In (fig. A.4) it is shown that the serial connection between the models H1 and H2, which is realized by using

```
>> H = series(H1, H2);
```

or else

```
>> H = H2 * H1;
```

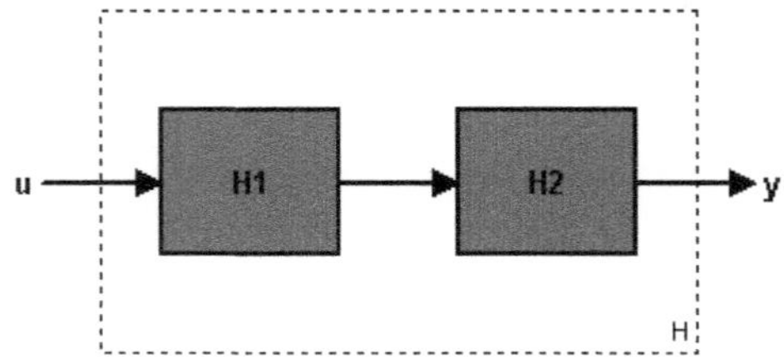

Fig. A.4 Serial connection configuration

In fig. A.5 it is shown that the parallel connection between the models H1 and H2, which is realized by using

```
>> H = parallel(H1, H2);
```

or else

```
>> H = H2 + H1;
```

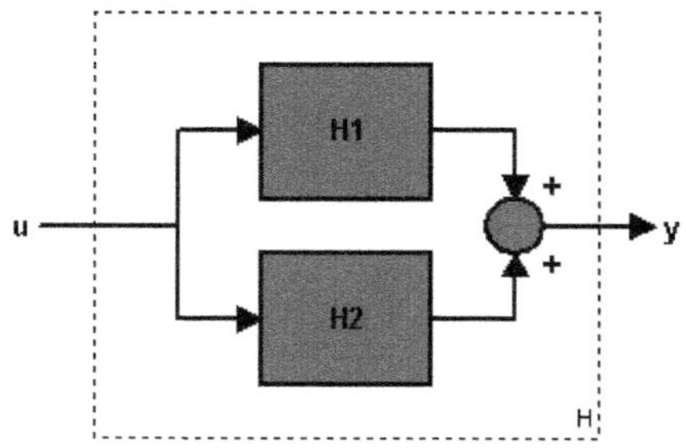

Fig. A.5 Parallel connection configuration

In fig. A.6 it is shown that closed loop configuration between the models H1 and H2 with negative feedback, which is realized by using

```
>> H = feedback (H1, H2);
```

The negative feedback is the default option, and if positive feedback is desired, a '+' sign must be specified as

```
>> H = feedback (H1, H2,+1);
```

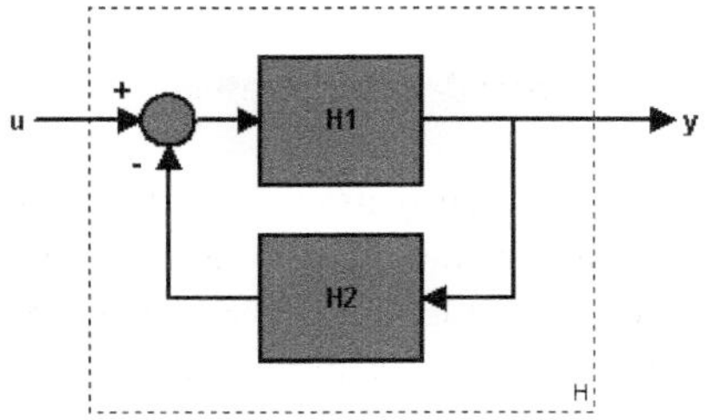

Fig. A.6 Closes loop connection configuration

The concatenation of models can be made in different ways (fig. A.7), (fig. A.8), (fig. A.9):

```
>> H = [H1,H2];
```

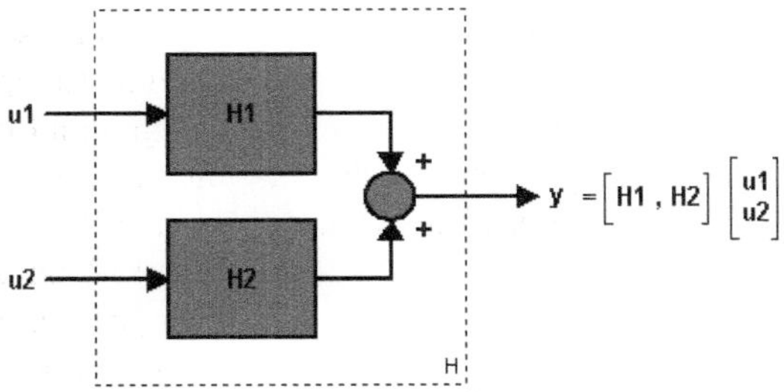

Fig. A.7 Concatenation of models (1 output, 2 inputs)

```
>> H = [H1; H2];
```

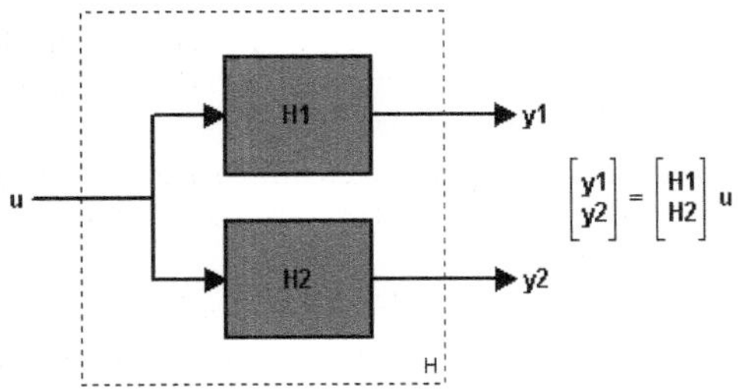

Fig. A.8 Concatenation of models (2 outputs, 1 input)

```
>> H = append(H1,H2);
```

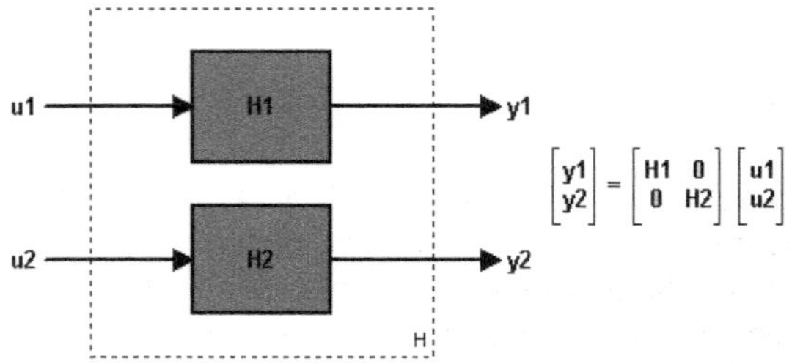

Fig. A.9 Concatenation of models (2 outputs, 2 inputs)

A.7 Conversion between Models

LTI models can be converted from one type to another (fig. A.10) by using the *tf*, *zpk* and *ss* commands,

```
>> System = tf (system);          % TF model converts
>> System = ZPK (system);      % ZPK model converts
>> System = ss (system);            % converts the SS model
```

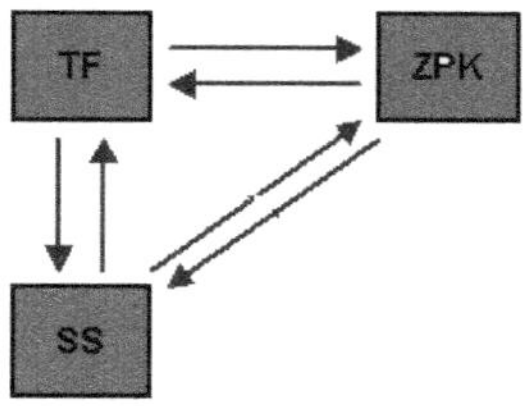

Fig. A.10 Conversion between model formats

Example

```
>> Gs= ss (-2,1,1,3);
```

can be converted to zeros, poles, and gain using

```
>> Gz = zpk(Gs)

Zero/pole/gain:
3 (s+2.333)
-----------
   (s+2)
```

A.8 Time Response

In order to compare the performance of different control systems, different types of standard test signals are used. This performance is analyzed by using response characteristics such as settling time, rise time, etc. Besides, the design specifications are often given as a function of the system's response to test signals such as step, ramp, or parabolic unit impulse. MATLAB incorporates these test signals, generating also the system response graphs.

The step response of a system can be obtained by application of the *step* command

```
>> step(sys,t)
```

where $t = t_0{:}\,\Delta t{:}\,t_f$ is the time horizon, with t_0, Δt and t_f being the initial, step and final time. Also it is possible to save the step response by using

```
>> [y,t,x] = step(sys)
```

The impulse response can be elicited by application of the *impulse* command

```
>> impulse(sys,t)
>> [y,t,x] = impulse(sys)
```

In case of a general input signal $u(t), t = t_o{:}\,\Delta t{:}\,t_f$ the system response is obtained by using the *lsim* command, so that

```
>> lsim(sys,u,t)
>>[y,t,x] = lsim(sys,u,t)
```

Example

Draw the step response of the system whose function transfer is given by $G(s) = \frac{10}{(s+2)(s+5)}$ for $0 < t < 5$.

We apply the following list of commands:

```
>> s=zpk('s');
>> G=10/((s+2)*(s+5));
>> t=0:0.1:5;
>> step(G,t)
>> grid
```

The step response is plotted in fig. A.11.

Example

Draw the response of the system whose function transfer is given by $G(s) = \frac{5}{s^3+2s^2+s+1}$ for an input signal $u(t) = 2e^{-t}\sin(\pi t),\ 0 < t < 10$.

We apply the following list of commands,

```
>> s=tf('s');
>> G=5/(s^3+2*s^2+s+1);
>> t=0:0.2:30;
```

```
>> u=2*exp(-t).*sin(pi*t);
>> lsim(G,u,t)
>> grid
```

The time response for $u(t)$is plotted in fig. A.12.

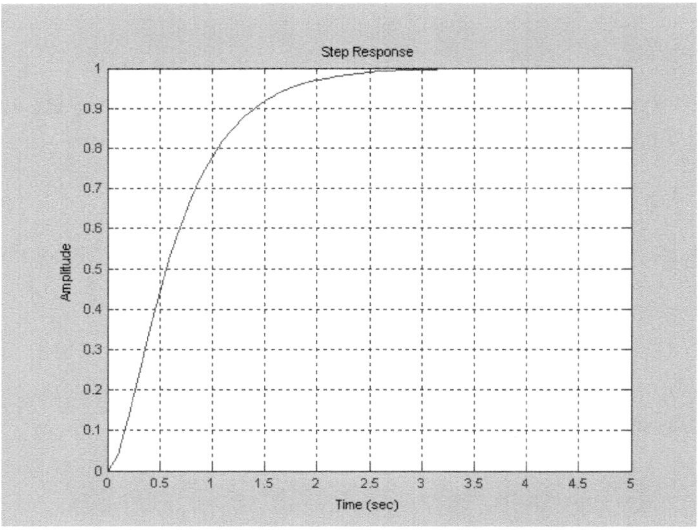

Fig. A.11 Step response of $G(s) = \frac{10}{(s+2)(s+5)}$

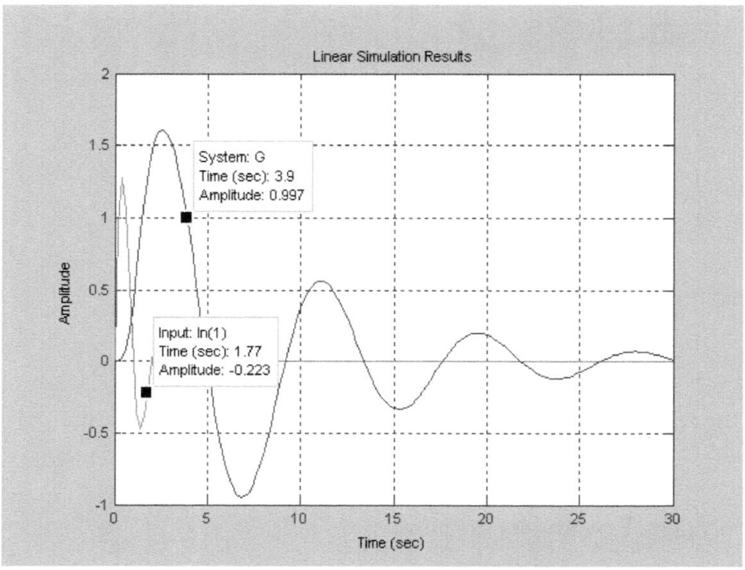

Fig. A.12 Time response for $G(s) = \frac{5}{s^3+2s^2+s+1}$.

Besides this, MATLAB can help to accomplish the partial fraction expansion of a rational function through the residues calculation.

The command *residue* finds the residues, poles and direct term of a partial fraction expansion of the ratio of two polynomials $C(s) = \frac{N(s)}{D(s)}$ according to

$$C(s) = \frac{N(s)}{D(s)} = \frac{K_1}{(s-p_1)} + \frac{K_2}{(s-p_2)} + \cdots + \frac{K_n}{(s-p_n)} + T(s)$$

by applying

```
>> [K,P,T] = residue(N,D);
```

Example

Find the poles and residues of $(s) = \frac{2s+1}{s^4+3s^3-7s+3}$.

```
>> s=tf('s');
>> C=(2*s+1)/(s^4+3*s^3-7*s+3);
>> [N,D]=tfdata(C,'v');
>> [K,P,T] = residue(N,D)

K =

  -0.0270 + 0.1888i
  -0.0270 - 0.1888i
   0.5000
  -0.4460


P =

  -2.2428 + 1.0715i
  -2.2428 - 1.0715i
   1.0000
   0.4856


T =

       []
```

Appendix B: The *rltool* Interactive Tutorial

The *rltool (*'Root Locus Tool') is a MATLAB guided user interface (GUI) used to perform the root locus analysis of linear and invariant single-input single-output systems.

This application provides a useful tool to realize both the design and the testing of controllers by using the root locus drawn. In this way, it is possible to change the gain or to add poles/zeros and see directly the results by viewing the system response when closed loop poles are moved throughout its root locus.

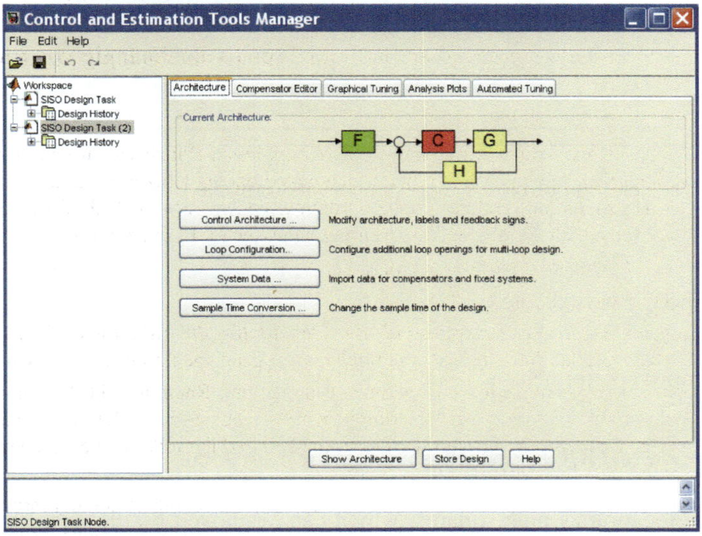

Fig. B.1 The Control and Estimation Tool Manager of *rltool.*

The rltool can be executed by typing

```
>> rltool
```

from the MATLAB command window or else

```
>> rltool(G)
```

in case we have created a transfer function ‘G’ previously. In both cases, the ‘Control and Estimation Tool Manager’ is activated, showing the closed loop configuration actually used, where ‘F’, ‘C’, ‘G’ and ‘H’ stand for filter, controller, plant and sensor, all of these represented by linear and invariant transfer functions (fig. B.1)

These transfer functions can be imported from the MATLAB command window if they are previously created by selecting the ‘System Data’ option (fig. B.2).

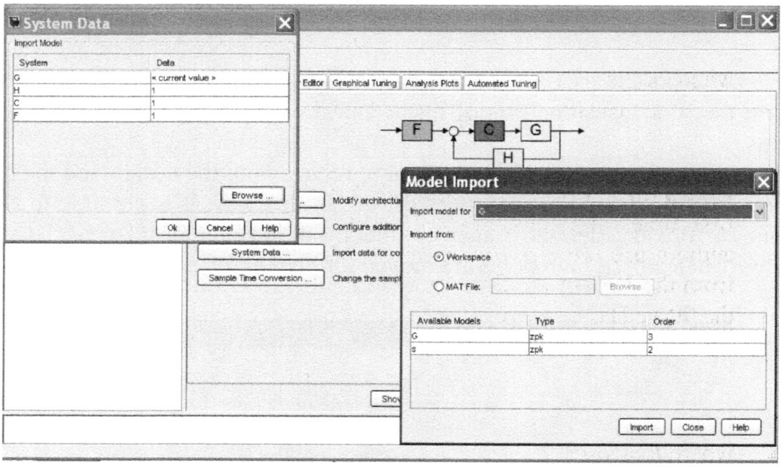

Fig. B.2 The selection of closed loop components form the Control and Estimation Tool Manager’

For instance, if we execute

```
>> s=zpk('s');
>> G= 5*(s+2)/(s*(s+1)*(s+5));
>> rltool(G)
```

then, the *rltool* GUI is opened showing the root locus for ‘G’ for the case of controller compensator C = 1 (fig. B.3). In this chart we can observe the closed loop pole location as pink squares, which can be moved by dragging them along the root locus. Note that the controller gain automatically changes in the ’Current Compensator” window (fig. B.4).

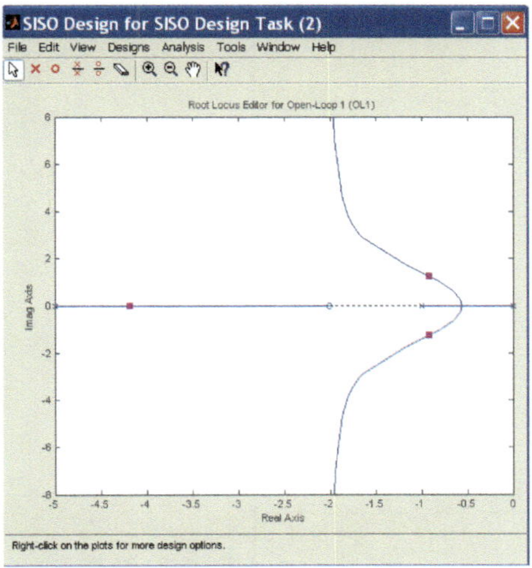

Fig. B.3 The window design of *rltool*, showing the root locus plot.

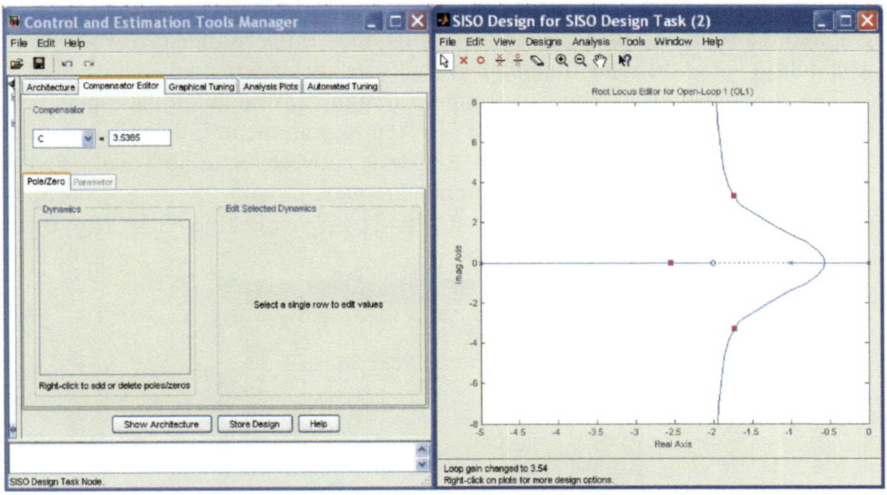

Fig. B.4 The Compensator Editor window of *rltool*.

This 'Compensation Editor' toolbar allows to add poles, add zeros or to delete either for the controller by selecting the 'Pole/Zero' option. This operation can also be made by right clicking and accessing to the 'Add Pole/Zero' or 'Delete Pole/Zero' option.

For instance, we can add a zero at s = -0.5 and an integrator at s = 0 we get another root locus plot with moving closed loop poles (fig. B.5).

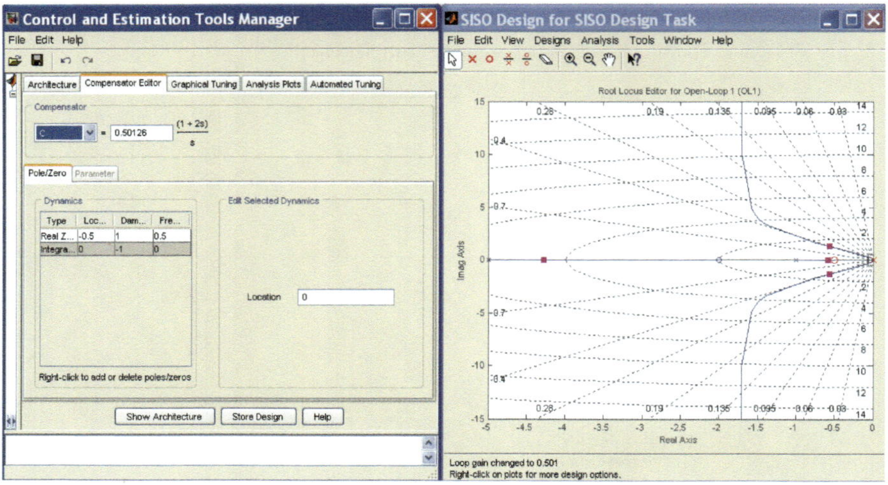

Fig. B.5 The Compensator Editor window of *rltool* showing the effect of added poles/zeros

It can be seen how this proportional integral controller has moved the root locus graph to the right half plane, by impairing its transient performance (even causing unstability) but at the same time improving its steady state regime.

The *rltool* also enables to super-impose design constraints on the s-plane. For this, we have to right click and select "Design Requirements", then "New" in the "Constraint type" menu. It is possible to choice between 'Settling Time', 'Percent Overshoot', 'Damping Ratio', 'Natural Frequency' and 'Region Constraint' requirements that can be viewed over an s-grid by right clicking 'Grid' (fig. B.6).

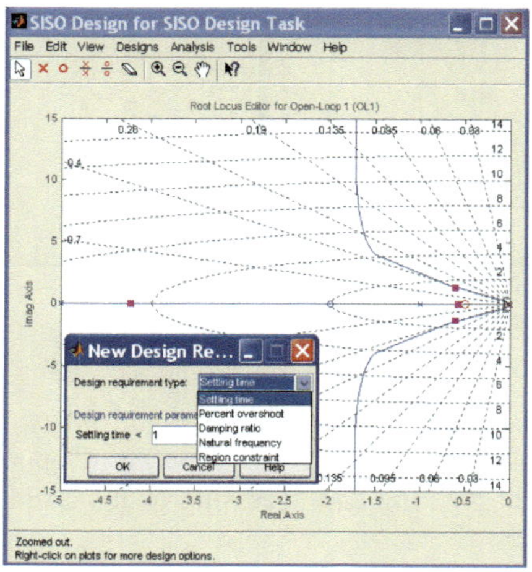

Fig. B.6 The Design Requirements setting of *rltool*

For example, we can select a settling time $t_{set} < 2\ s$ and a percent overshoot $M_p < 20\ (\%)$. It can be seen that these specifications cannot be met with this controller, no matter what we set the proportional gain to. That is, the two dominant roots of the closed loop system will never be in the valid region (fig. B.7).

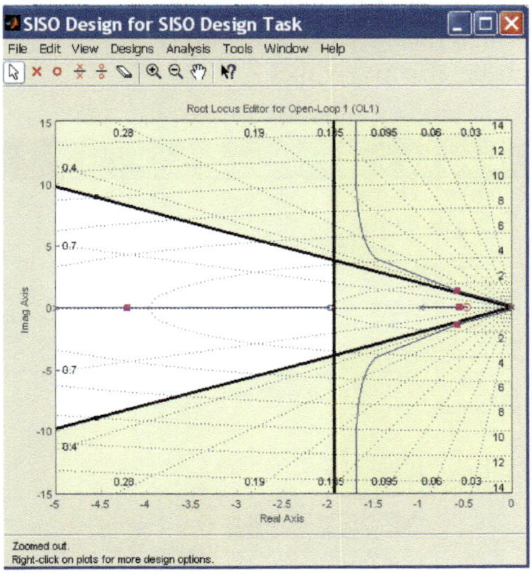

Fig. B.7 Imposing specific requirements in *rltool*

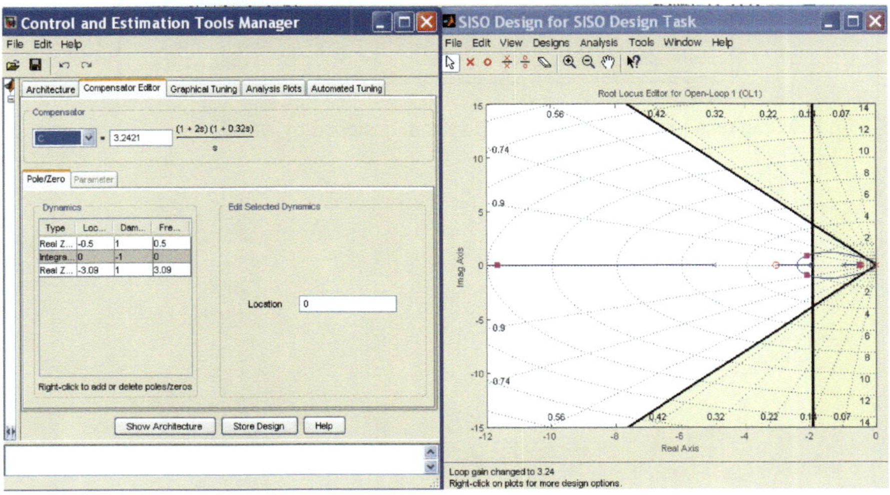

Fig. B.8 Fulfilling requirements by modifying the controller in *rltool*

Nevertheless, we can add a new the controller zero by using the toolbar in the *rltool* main window and drag one of the closed loop poles until they are inside the white region. Note how the new compensator transfer function designed meets the requirements in spite of the closed loop pole closer to the zero at s= -0.5 due to the cancellation effects (fig. B.8).

Besides we can see the time responses of the closed loop system by selecting the 'Analysis' at the toolbar and the 'Response to Step Command option. Then, a new plot window is opened, showing the step response from the reference to the system's output (fig. B.9). By right clicking, it can be shown that the time response characteristics (peak response, settling time, etc.) in order to test if the requirements have been effectively met.

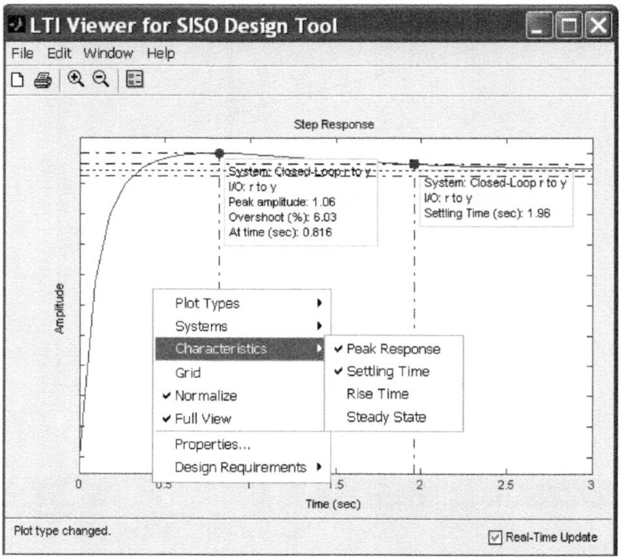

Fig. B.9 Step response of the closed loop system designed by *rltool*

Appendix C:
The SIMULINK Interactive Tutorial

C.1 Introduction

SIMULINK is an extension to MATLAB that allows users to rapidly and accurately build computer models of dynamical systems using block diagrams.

A block diagram describes a set of relationships that holds simultaneously, so the block diagram can be thought of being represented by a set of simultaneous equations.

The block diagram is assembled by gathering the constitutive blocks from the SIMULINK Library Browser, which are grouped by categories into different libraries (fig. C.1).

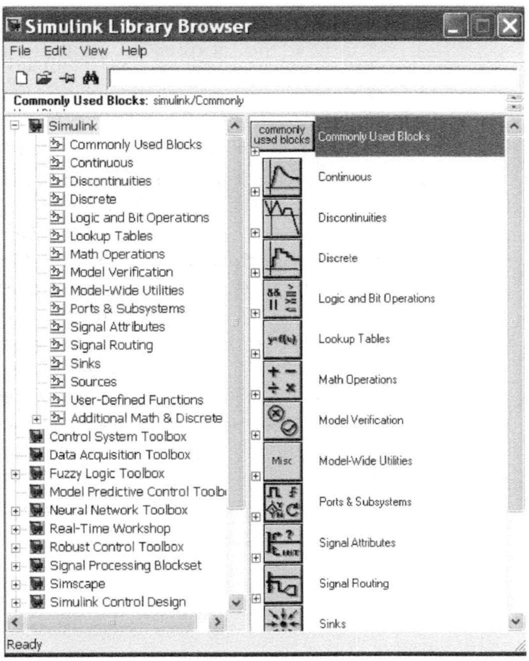

Fig. C.1 The SIMULINK Library browser

In order to access SIMULINK we can type in the MATLAB command window

```
>> SIMULINK
```

or else click on the SIMULINK icon (fig. C.2)

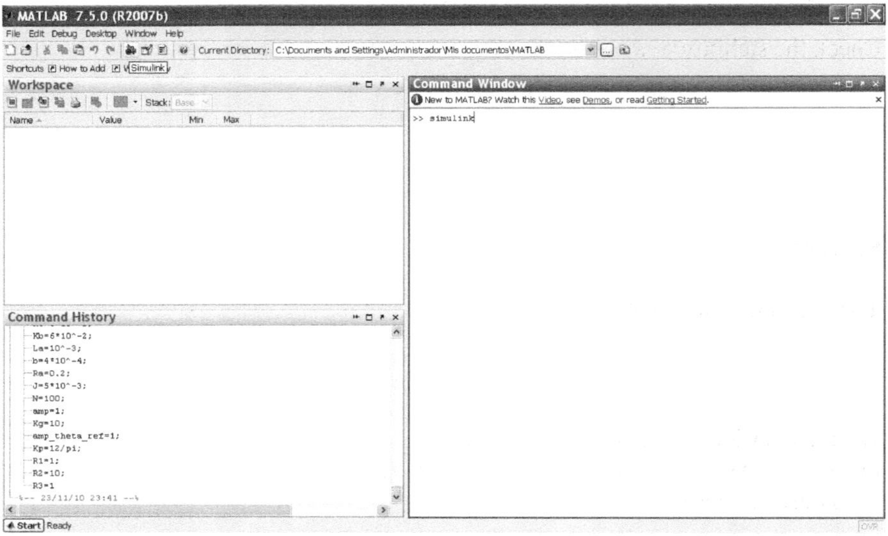

Fig. C.2 The access to SIMULINK from MATLAB

Once we access SIMULINK a new model must be created by dragging blocks from the SIMULINK Library browser to this new model window, by selecting the appropriate library where the searched block is included.

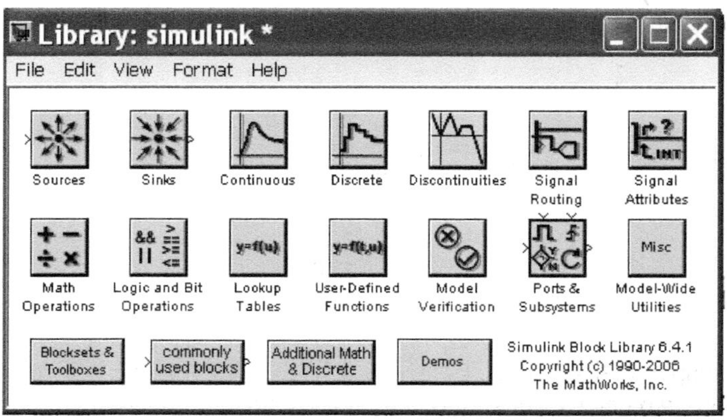

Fig. C.3 The SIMULINK Block Library

For the purpose of modeling, simulation, and control of dynamic systems, a subset of SIMULINK libraries will be used, namely, Continuous, Discontinuities, Lookup

Tables, Math Operations, Ports and Subsystems, Signal Routing, Sinks, Sources, User-Defined Functions and Additional Linear (included into SIMULINK Extras) (fig. C.3)

C.2 Creating a Model

In order to explain the process followed to create a SIMULINK model, we have selected a simple feedback control system for a ship (fig. C.4).

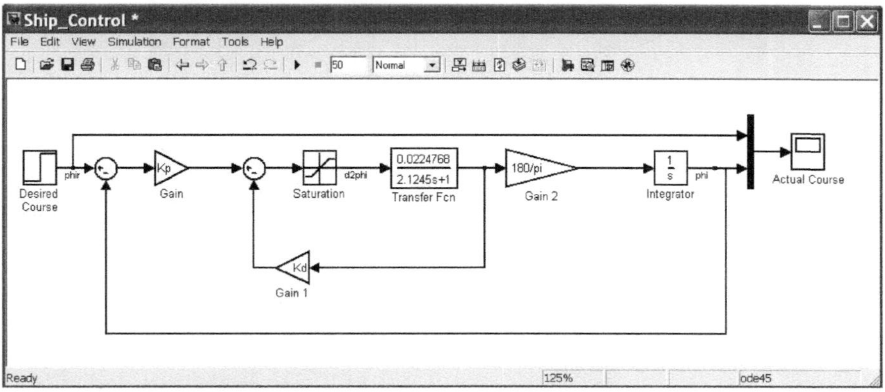

Fig. C.4 Feedback control system for a ship

In the first place we have to collect all blocks necessary to build the system from the SIMULINK block Library. Afterwards, we will connect the selected blocks to implement the block diagram, also changing the parameters of these blocks if it is necessary. Finally the system is simulated by selecting the appropriate solver options.

Therefore, as first step, we open a new model, dragging the constitutive blocks of the block diagram from each one of the block libraries where they are included (fig. C.5)

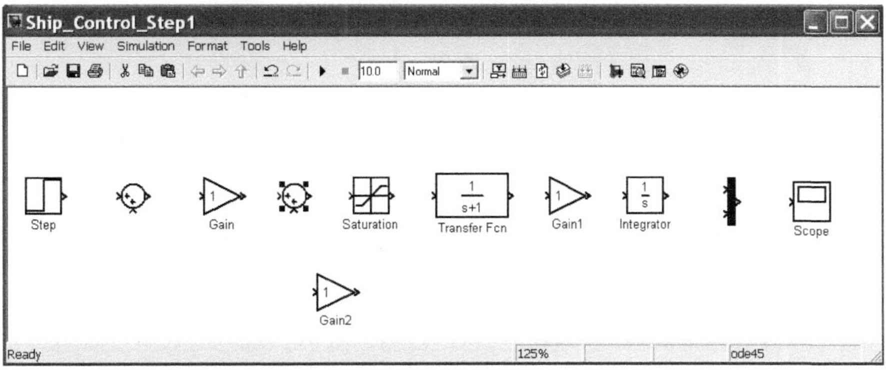

Fig. C.5 Building process for the feedback control of a ship (step 1)

As second step, we connect these blocks following the desired connection pattern as indicated in the feedback control system (fig. C.6). Possibly, it will be necessary to flip or rotate some blocks or to derive some pickoff points.

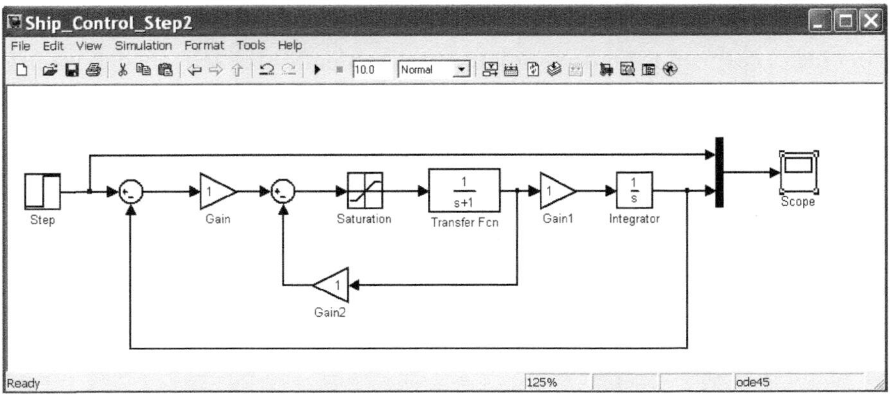

Fig. C.6 Building process for the feedback control of a ship (step 2)

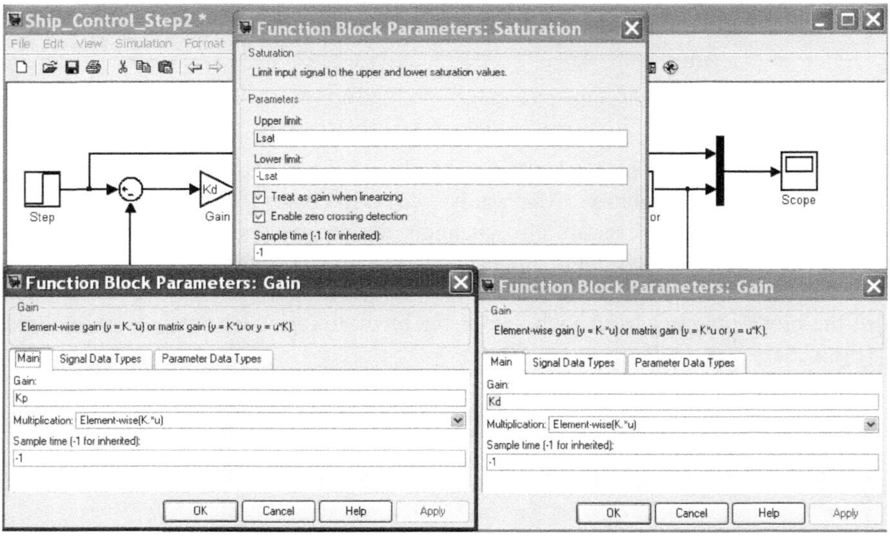

Fig. C.7 Building process for the feedback control of a ship (step 3)

Afterword, in a third step we have to change the block's parameters according to the values defined in the original feedback control system if it is necessary, adding optionally labels both to identify the signals transferred between blocks or to name specific blocks (fig. C.7). Block parameters can also be defined from the MATLAB workspace and used in SIMULINK. In this case we have assigned the gains, maxim saturation, and reference step as

```
>> Kd=1.8;
>> Kp=1.5;
```

```
>> Lsat=10;
>> phir=90;
```

Before simulation is run, we must select the appropriate simulation time and the numerical method used, including the integration step value in case of fixed-step selection (fig. C.8).

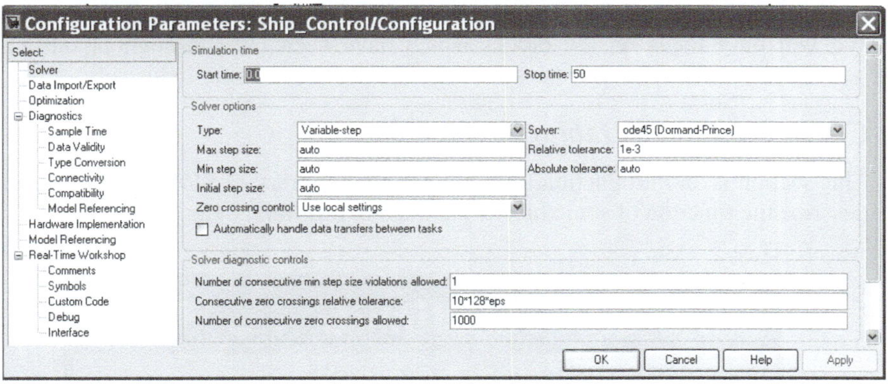

Fig. C.8 Definition of the solver configuration for the feedback control.

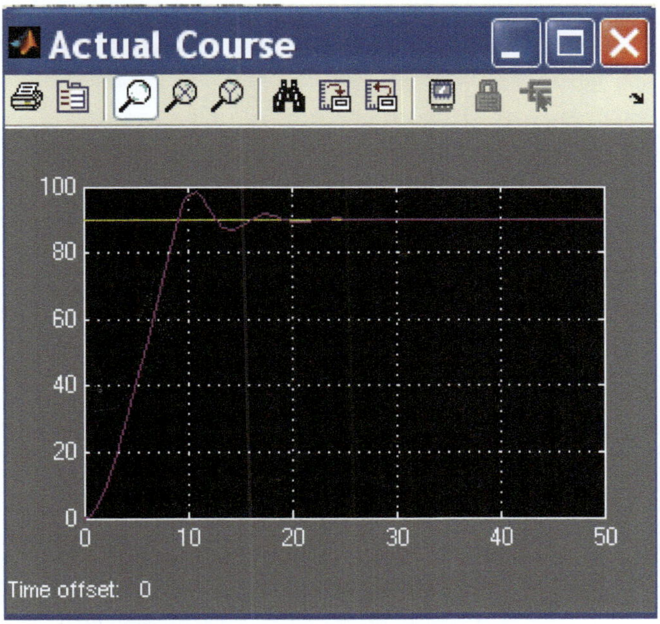

Fig. C.9 Step response of the DC motor obtained by using SIMULINK

Finally, the simulation is run by double-clicking on the ▻ icon and output responses can be viewed by double-clicking on the Scope block to view its output. Hit the auto-scale button to see the step response of the feedback controller applied to a reference course for a step amplitude of 90° (fig. C.9)

C.3 SIMULINK Libraries

SIMULINK contains a large number of blocks from which models can be built. These blocks are arranged in Block Libraries which are accessed by the main SIMULINK window.

The set of Block Libraries we are going to describe contains the main blocks used for analysis, modeling and control of dynamic system, nonlinear in general. We will only focus on the blocks which have been used throughout the text, leaving aside other blocks less used in the System Engineering field.

C.3.1 Continuous Library

This section is an introduction to the Continuous Library of SIMULINK. We will describe the function of some blocks included in this library (fig. C.10)

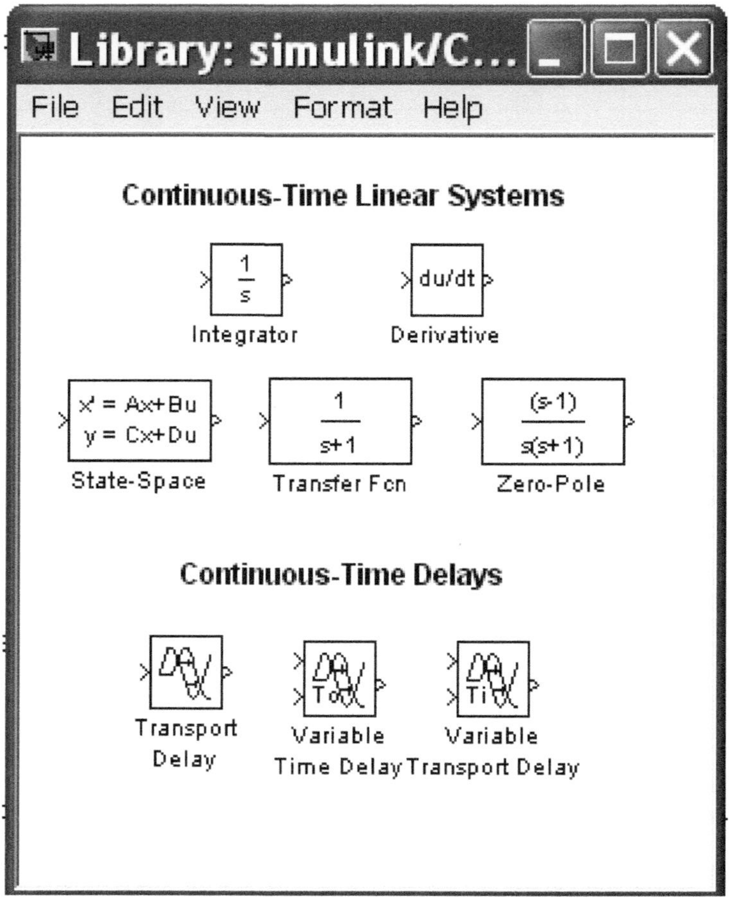

Fig. C.10 The SIMULINK Continuous Library

Integrator

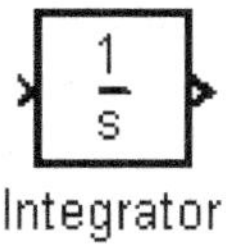

The Integrator block integrates its input and is used with continuous–time signals. We can use different numerical integration methods to compute the Integrator block's output.

Transfer Function

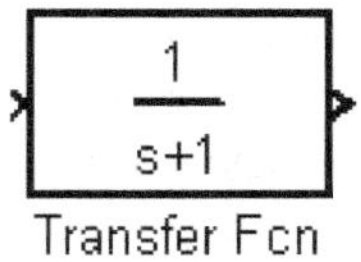

The **Transfer Fcn** block implements a transfer function as a ratio of polynomials.

Zero-Pole

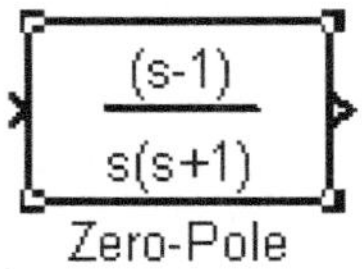

The **Zero-Pole** block implements a system with the specified zeros, poles, and gain in the s domain.

State-Space

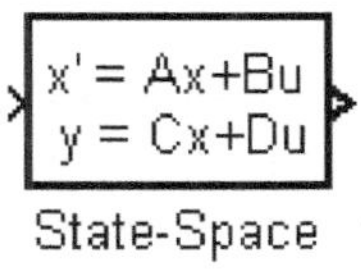

The **State-Space** block implements a system defined by the state-space equations.

Derivative

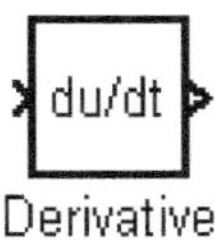

The **Derivative block** approximates the derivative of its input. The initial output for the block is zero. The accuracy of the results depends on the size of the time steps taken in the simulation.

Transport Delay

The **Transport Delay** block delays the input by a specified amount of time. It can be used to simulate a time delay.

C.3.2 Source Library

This section is an introduction to the Source Library of SIMULINK. We will describe the function of some blocks included in this library (fig. C.11).

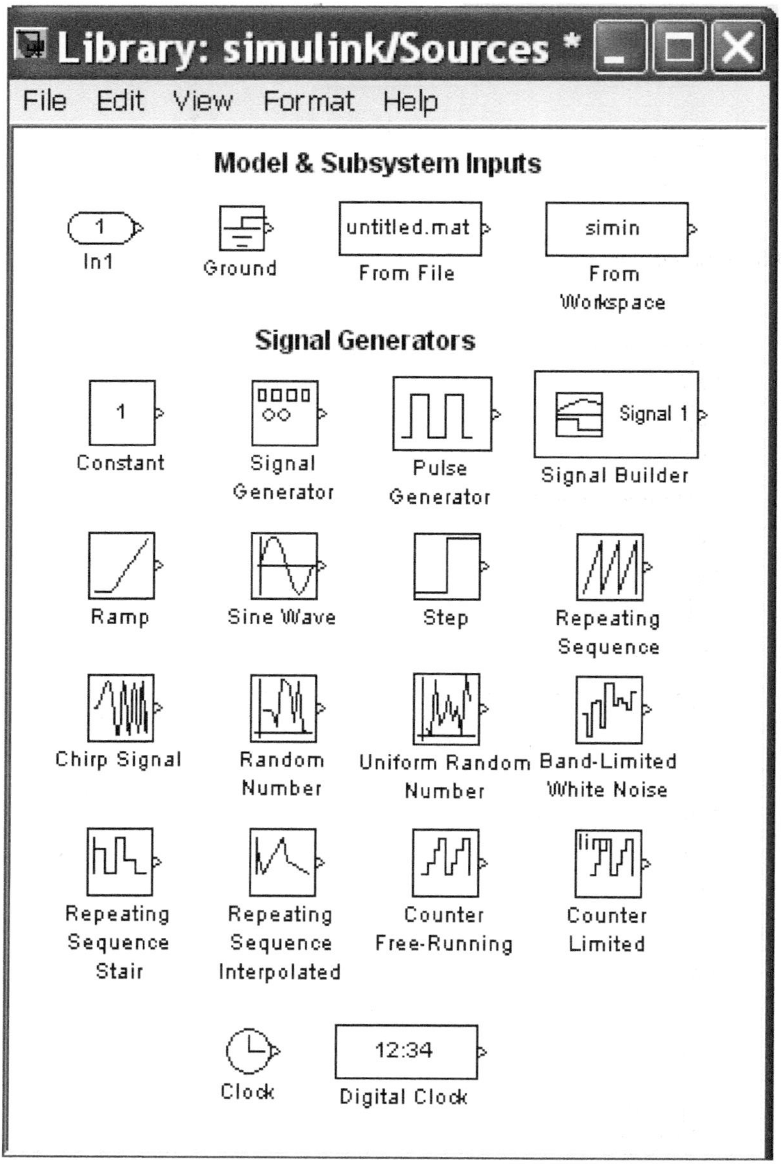

Fig. C.11 The SIMULINK Source Library

Constant block

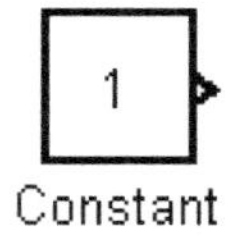

The **Constant block** is used to define a real or complex constant value. This block accepts scalar, vector, or matrix.

Step

The **Step** block generates a step between two defined levels at some specified time.

Ramp

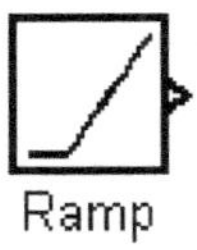

The **Ramp** block generates a signal that starts at a specified time and value and changes by a specified rate. The characteristics of the generated signal are determined by the specified Slope and Start time.

Sine Wave

The **Sine Wave** block generates a sine wave. To generate a cosine wave, we specify the Phase parameter as $\frac{\pi}{2}$.

Pulse Generator

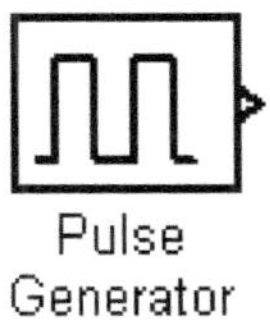

The **Pulse Generator** block generates square wave pulses at regular intervals. The shape of the generated waveform depends on the parameters, Amplitude, Pulse Width, Period, and Phase Delay as shown in (fig. C.12).

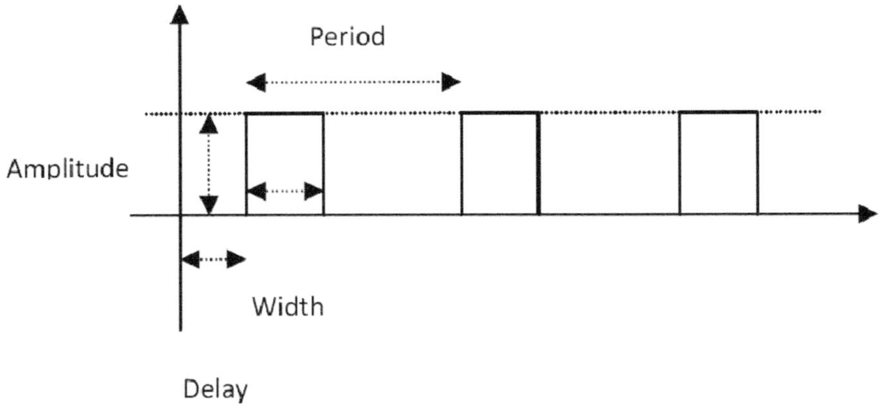

Fig. C.12 Illustration of the Pulse Generator block parameters

Signal Builder

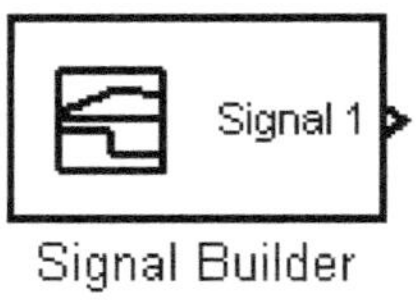

The **Signal Builder** block allows the user to create interchangeable groups of piece-wise linear signal sources.

Digital Clock

The **Digital Clock** block displays the simulation time at a specified sampling interval.

From File

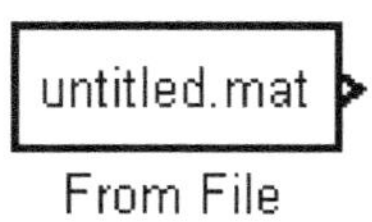

The **From File** block outputs data read from a MAT file. The name of the file is displayed inside the block.

From Workspace

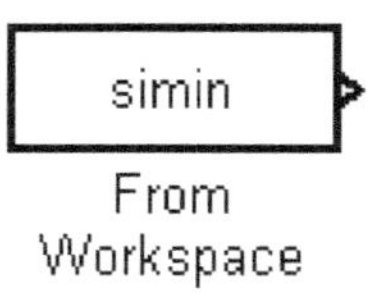

The **From Workspace** block reads data from the MATLAB workspace.

Signal Generator

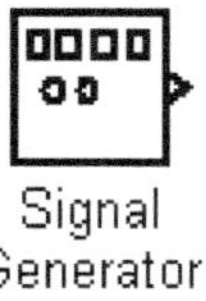

The **Signal Generator** block can produce one of four different waveforms: sine wave, square wave, saw tooth wave, and random wave.

C.3.3 Sinks Library

This section is an introduction to the Sinks Library of SIMULINK. We will describe the function of some blocks included in this library (fig. C.13).

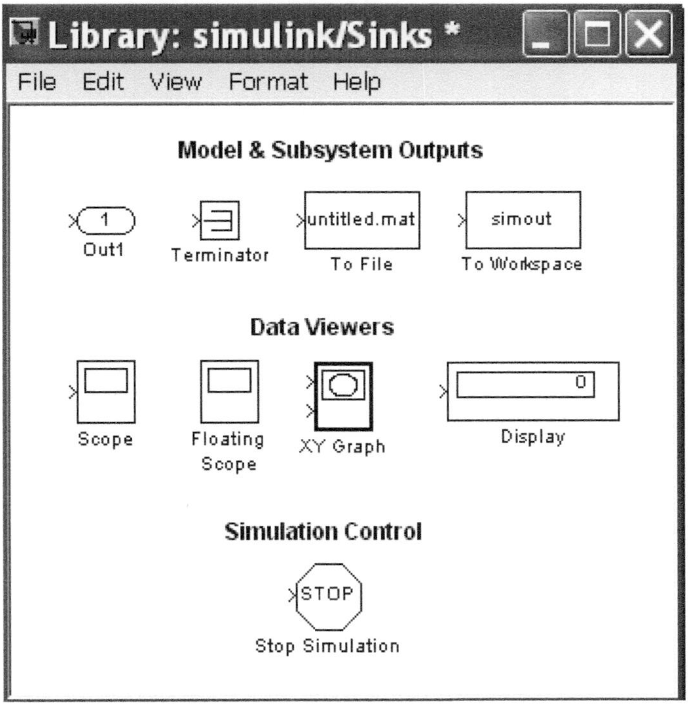

Fig. C.13 The SIMULINK Sinks Library

Scope

The **Scope** block displays its input with respect to simulation time. The Scope block can have multiple axes (one per port), but all axes have a common time range with independent y–axes. We can move and resize the Scope window and we can modify the Scope's parameter values during the simulation.

Display

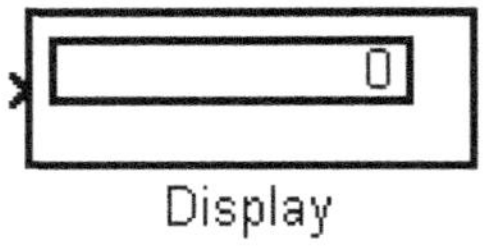

The **Display** block shows the value of its input on its icon. The display formats are the same as those of MATLAB.

XY Graph

The **XY Graph** block displays an X–Y plot of its inputs in a MATLAB figure window. This block plots data in the first input (the x direction) against data in the second input (the y direction).

To File

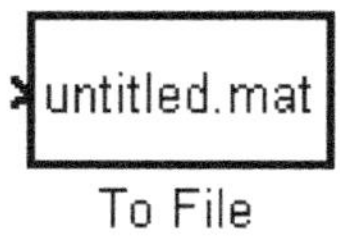

The **To File** block writes its input to a matrix in a MAT–file. The block writes one column for each time step: the first row is the simulation time; the remainder of the column is the input data.

To Workspace

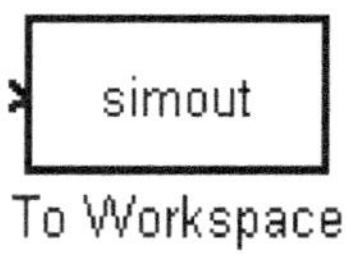

The **To Workspace** block writes its input to the workspace. This block writes its output to an array or structure that has the name specified by the block's Variable name parameter. The Save format parameter determines the output format.

Stop Simulation

The Stop Simulation blocks the simulation when the input signal is "TRUE".

C.3.4 Math Operations Library

This section is an introduction to the Math Operations Library of SIMULINK. We will describe the function of some blocks included in this library (fig. C.14).

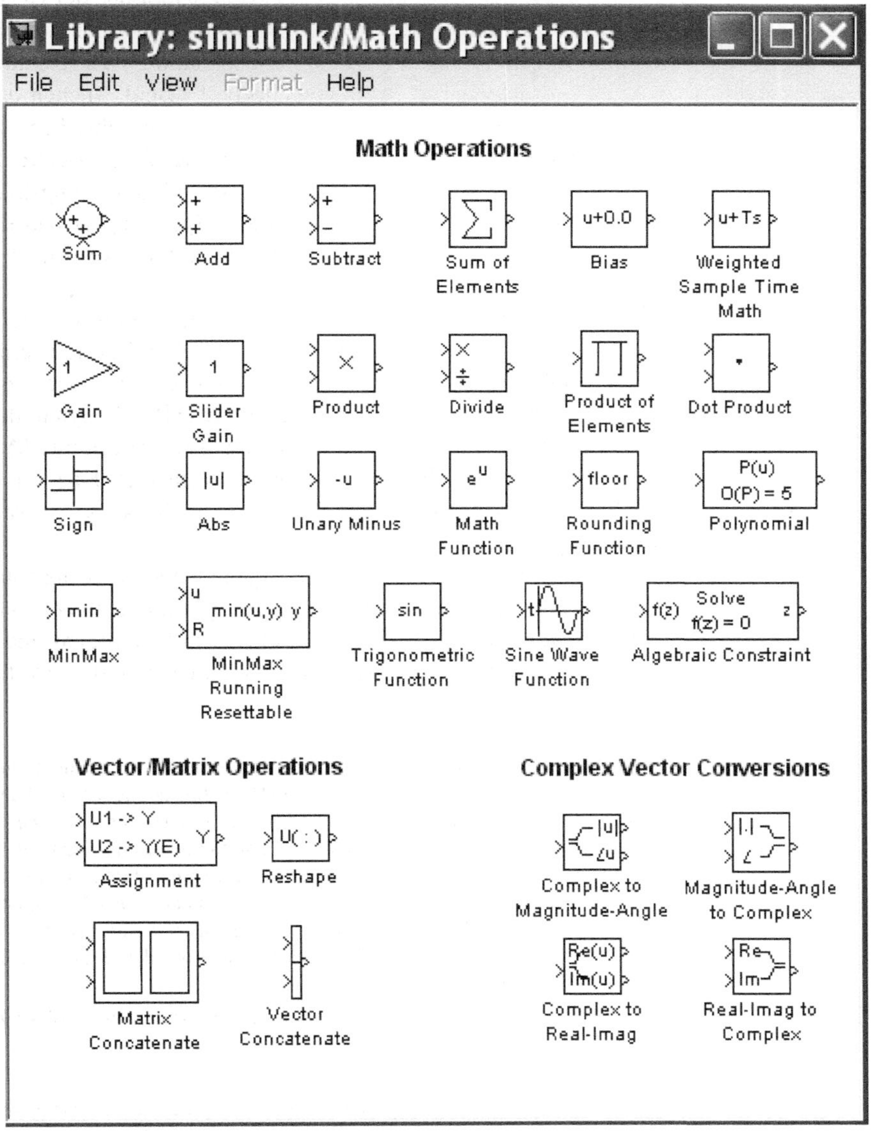

Fig. C.14 The SIMULINK Math Operations Library

Sum

The **Sum** block is an implementation of the Add block which is described below. We can choose the icon shape (round or rectangular) of the block on the Block Parameters dialog box.

Add

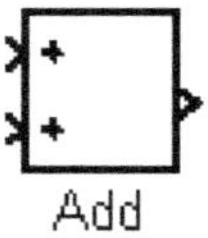

The **Add** block performs addition or subtraction on its inputs. This block can add or subtract scalar, vector, or matrix inputs. We specify the operations of the block with the list of sign parameters, plus (+), minus (-).

Gain

The **Gain block** multiplies the input by a constant value (gain). The input and the gain can each be a scalar, vector, or matrix.

Product block

The **Product block** performs multiplication or division of its inputs. We specify the operations with the number of inputs parameter. Multiply (*) and divide (/) characters indicate the operations to be performed on the inputs.

Divide

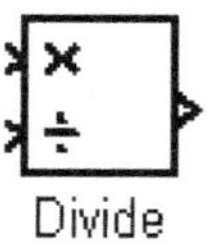

The **Divide block** is an implementation of the Product block. It can be used to multiply or divide inputs.

Sign

The **Sign** block indicates the sign of the input. The output is 1 when the input is greater than zero, the output is 0 when the input is equal to zero, and the output is −1 when the input is less than zero.

Abs

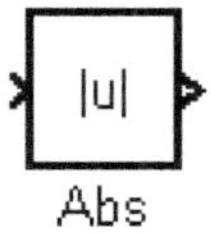

The **Abs** block outputs the absolute value of the input.

Math Function

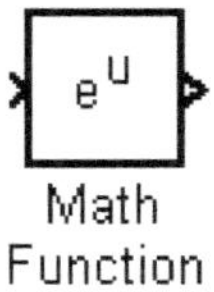

The **Math Function** block performs the following mathematical functions: exp, log, 10^u, log10, magnitude^2, square, sqrt, pow, etc.

MinMax

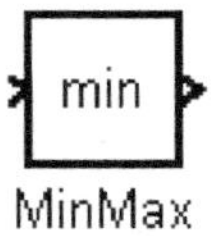

The **MinMax** block outputs either the minimum or the maximum element or elements of the inputs. We choose the function to apply by selecting one of the choices from the Function parameter.

Trigonometric Function

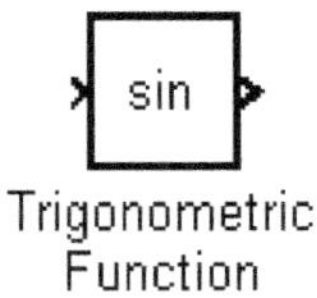

The **Trigonometric Function** block performs the trigonometric functions sin, cos, tan, asin, acos, atan, and the hyperbolic functions sinh, cosh, tanh, asinh, acosh, and atanh.

C.3.5 Lookup Table Library

This section is an introduction to the Lookup Table Library of SIMULINK. We will describe the function of some blocks included in this library (fig. C.15).

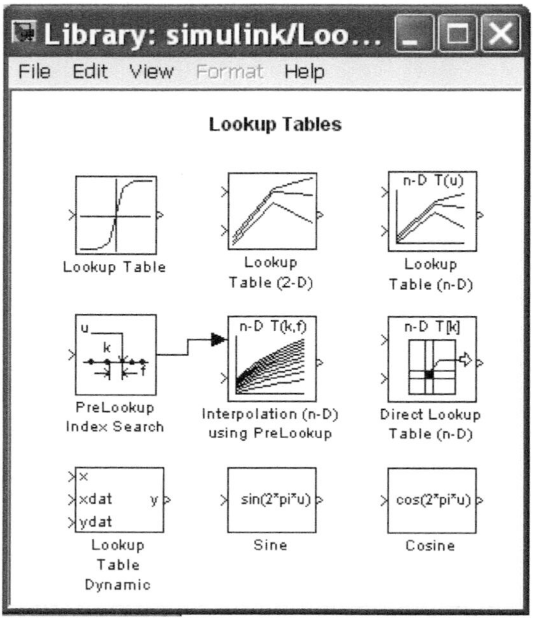

Fig. C.15 The SIMULINK Lookup Tables Library

Lookup Table

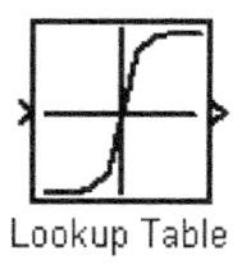

The **Lookup Table** block computes an approximation to a function $y = f(x)$ where the data vectors x and y are given, and it is required that the x data vector must be monotonically increasing.

Lookup Table (2–D)

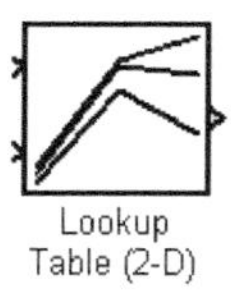

The **Lookup Table (2–D)** block computes an approximation for a function $z = f(x, y)$ when the data points x, y, and z are given.

Lookup Table (n–D)

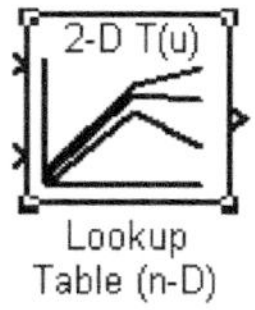

The **Lookup Table (n–D)** block n–dimensional interpolated table lookup including index searches. The table is a sample representation of a function of N variables.

C.3.6 Discontinuities Library

This section is an introduction to the Discontinuities Library of SIMULINK. We will describe the function of some blocks included in this library (fig. C.16).

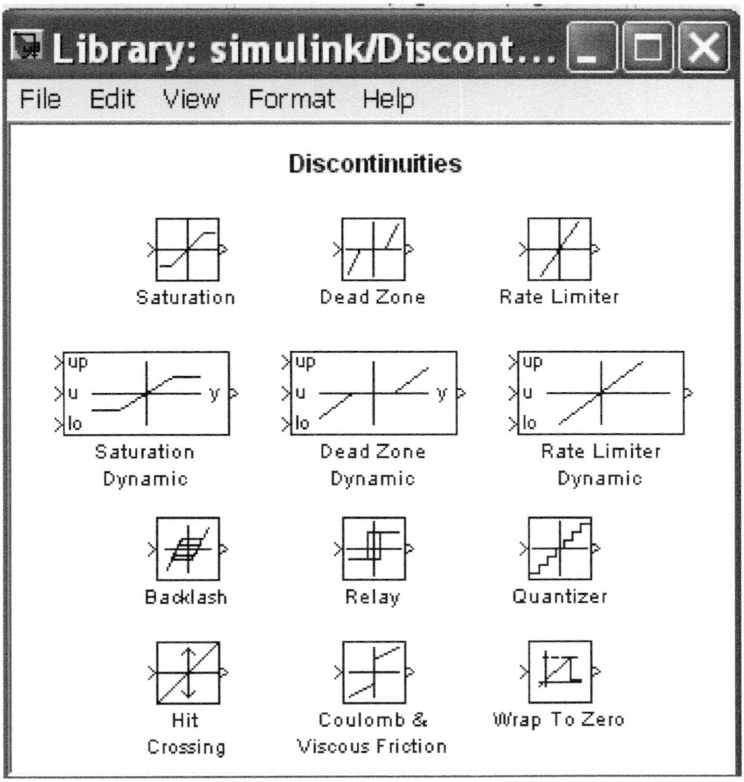

Fig. C.16 The SIMULINK Discontinuities Library

Saturation

The **Saturation** block sets upper and lower bounds on a signal. When the input signal is within the range specified by the Lower limit and Upper limit parameters, the input signal passes through unchanged.

Dead Zone

The **Dead Zone** block generates zero output within a specified region, called its **dead zone**. The lower and upper limits of the dead zone are specified as the Start of dead zone and End of dead zone parameters.

Rate Limiter

The **Rate Limiter** block limits the first derivative of the signal passing through it.

Backlash

The **Backlash** block implements a system in which a change in input causes an equal change in output. However, when the input changes direction, an initial change in input has no effect on the output.

Relay

The **Relay** block output can switch between two specified values. When the relay is on, it remains on until the input drops below the value of the Switch off point parameter. When the relay is off, it remains off until the input exceeds the value of the Switch on point parameter.

C.3.7 Signal Routing Library

This section is an introduction to the Signal Routing Library of SIMULINK. We will describe the function of some blocks included in this library (fig. C.17).

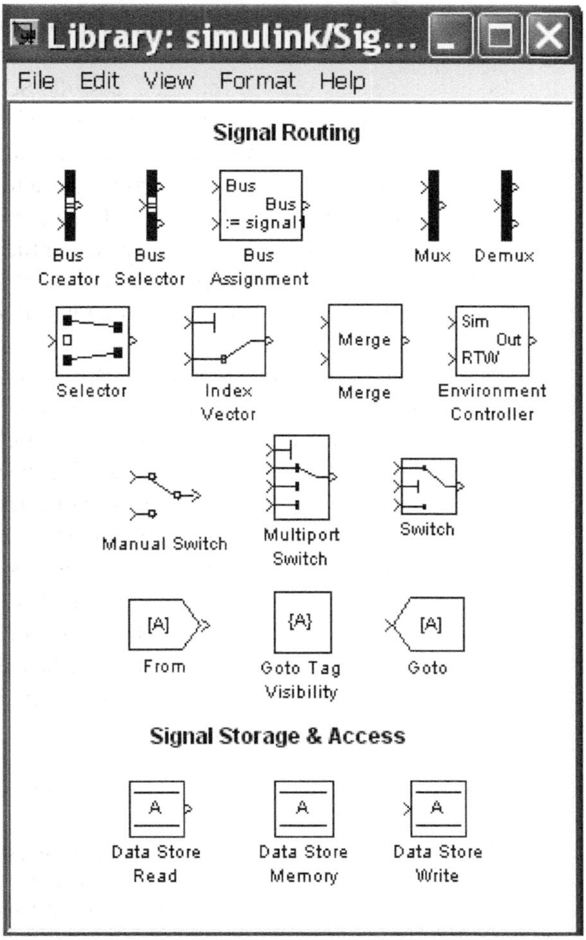

Fig. C.17 The SIMULINK Signal Routing Library

Mux

The **Mux** block combines its inputs into a single output. An input can be a scalar, vector, or matrix signal.

Demux

The **Demux** block extracts the components of an input signal and outputs the components as separate signals.

Switch

The **Switch** block outputs the first (top) input or the third (bottom) input depending on the value of the second (middle) input. The first and third inputs are the **data inputs**. The second input is the **control input**.

From

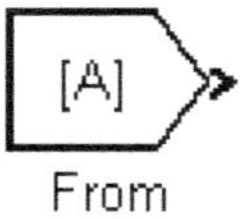

The **From** block accepts a signal from a corresponding Goto block , and passes it as output.

Goto

The **Goto** block passes its input to its corresponding From blocks. From and Goto blocks allow us to pass a signal from one block to another without actually connecting them.

C.3.8 Ports and Subsystems Library

This section is an introduction to the Ports and Subsystems Library of SIMULINK. We will describe the function of some blocks included in this library (fig. C.18).

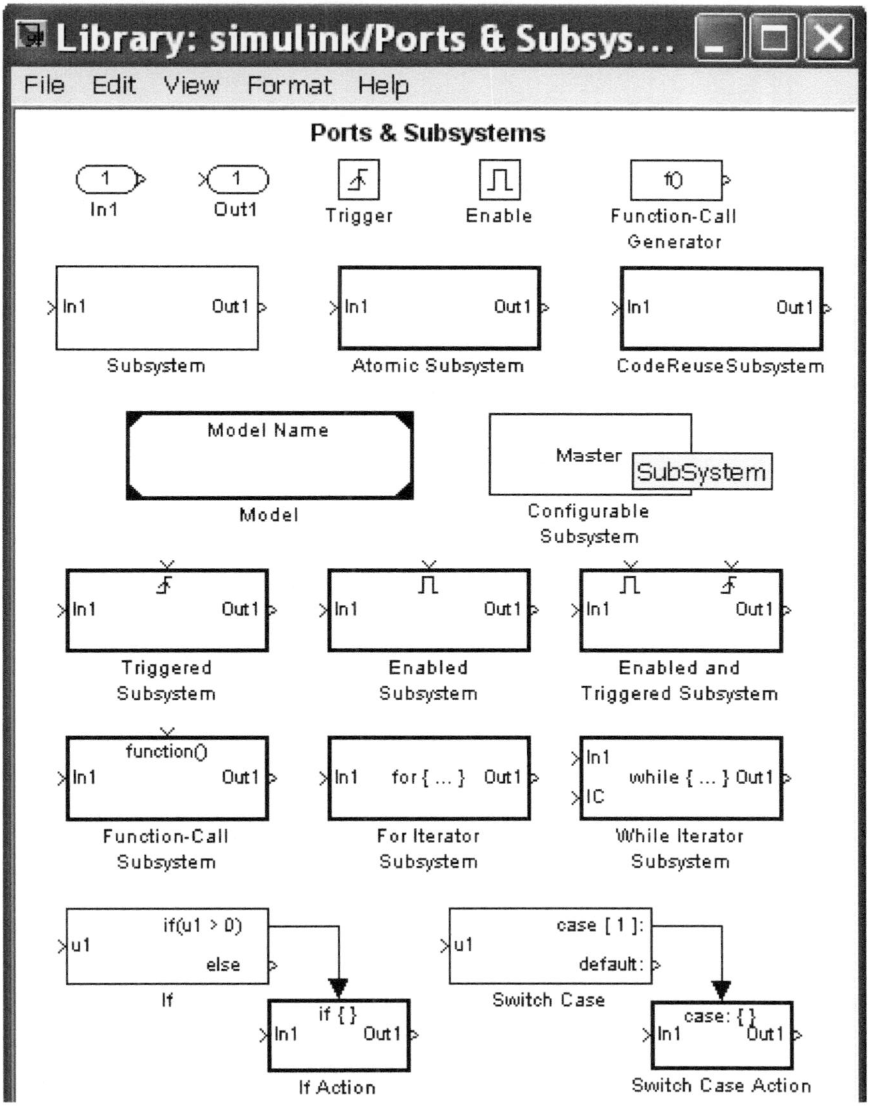

Fig. C.18 The SIMULINK Ports and Subsystems Library

Inport, Outport, and Subsystem

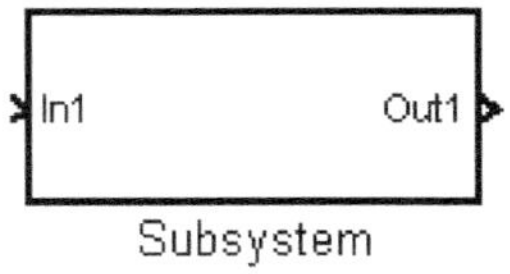

Inport blocks are ports that serve as links from outside a system into the system. **Outport blocks** are output ports for a subsystem. A **Subsystem block** represents a subsystem of the system that contains it. As our model increases in size and complexity, we can simplify it by grouping blocks into subsystems.

C.3.9 User-Defined Functions Library

This section is an introduction to the User-defined Functions Library of SIMULINK. We will describe the function of some blocks included in this library (fig. C.19).

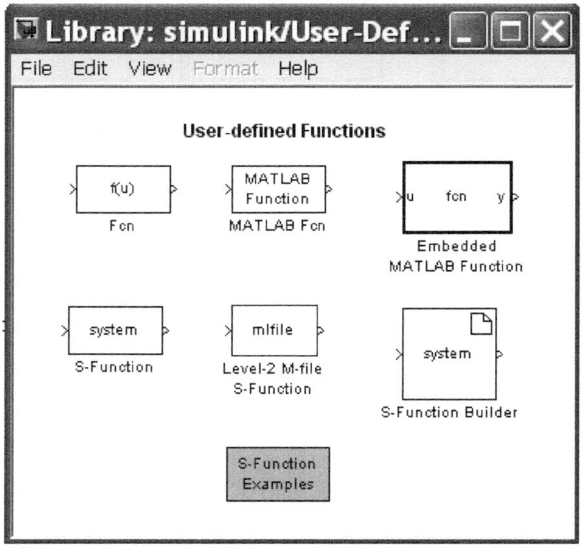

Fig. C.19 The SIMULINK User-defined Functions Library

Fcn

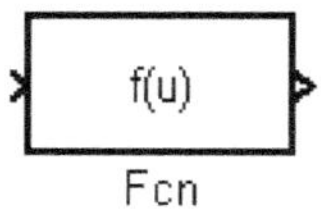

The **Fcn** block applies a specified expression to its input denoted as u. If u is a vector, u(i) represents the ith element of the vector; u(1) or u alone represents the first element. The specified expression can consist of numeric constants, arithmetic operators, relational operators, logical operators, and the math functions.

MATLAB Fcn

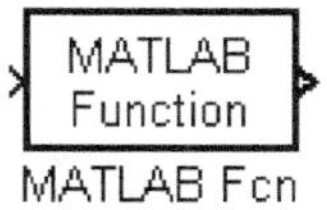

The **MATLAB Fcn** block applies the specified MATLAB function or expression to the input.

C.3.10 Additional Linear Library

This section is an introduction to the Additional Linear Library of SIMULINK, which is embedded into the SIMULINK Extras Library. We will describe the function of some blocks included in this library (fig. C.20).

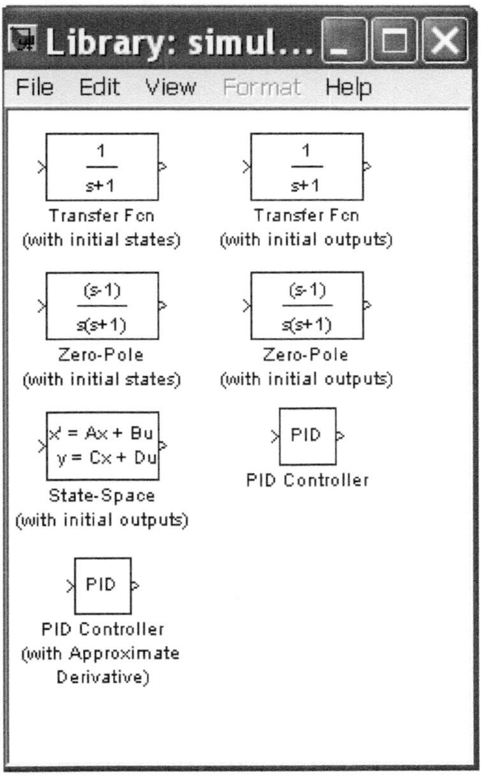

Fig. C.20 The SIMULINK Additional Linear Library

Transfer Fcn (with initial states)

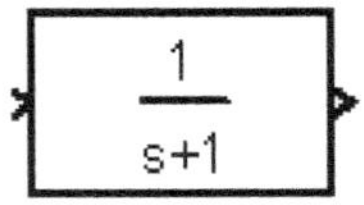

Transfer Fcn
(with initial states)

Transfer Fcn (with initial output)

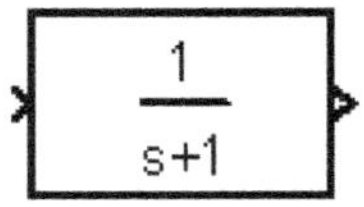

Transfer Fcn
(with initial outputs)

Zero-Pole (with initial states)

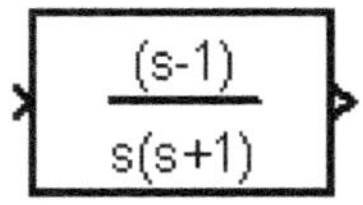

Zero-Pole
(with initial states)

Specify the initial output for the transfer function as a ratio of polynomials.

Zero-Pole (with initial ouput)

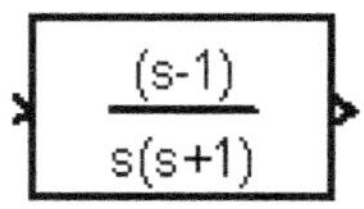

Zero-Pole
(with initial outputs)

Specify the initial output for the transfer function given by zero, pole, and gain format.

States-Space (with initial ouput)

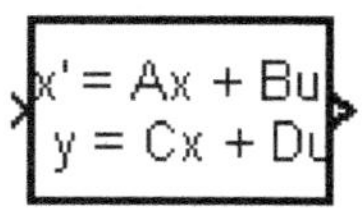

State-Space
(with initial outputs)

Specify the initial output for the system given in state space format.

PID Controller

PID Controller

Implements the proportional-integral-derivative action based on the proportional, integral and derivative gains.

PID Controller (with Approximate Derivative)

PID Controller
(with Approximate
Derivative)

Implements the proportional-integral-derivative action, with derivative term implemented using an s/(s/N +1) transfer function block.

Appendix D: The SIMSCAPE Modeling Environment Tutorial

D.1 Introduction

SIMSCAPE is a MATLAB-based, object-oriented physical modeling language that enables the user to create models of physical components using an acausal modeling approach.

This language is designed for use in the MATLAB and SIMULINK environments, for it can benefit MATLAB functions and SIMULINK blocks (fig. D.1).

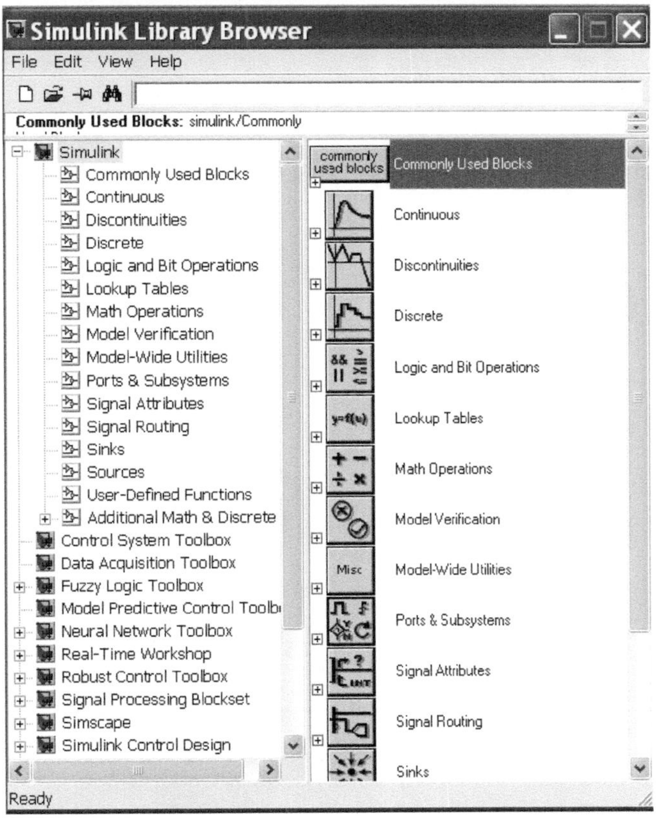

Fig. D.1 The SIMULINK Library browser

The integration of control systems with physical systems covering multiple physical domains such as electrical, mechanical, hydraulic, etc., requires a useful model representation of the physical system. The physical network approach enables the users to create models of physical components that can cover multiple physical domains and can also be reusable.

The SIMSCAPE methodology is based on the connection of physical components, each one with its dynamic equations embedded. The specific connection diagram together with the conservation laws applied, determines the system dynamic equations.

The physical component diagram is assembled by gathering the constitutive components from the Foundation Library of the SIMSCAPE Library Browser, which can be accessed from the SIMULINK Library Browser (fig. D.2). The SIMSCAPE components are grouped by categories into different libraries of components (Electrical, Hydraulic, Mechanical, Physical Signals, Thermal) and Utilities.

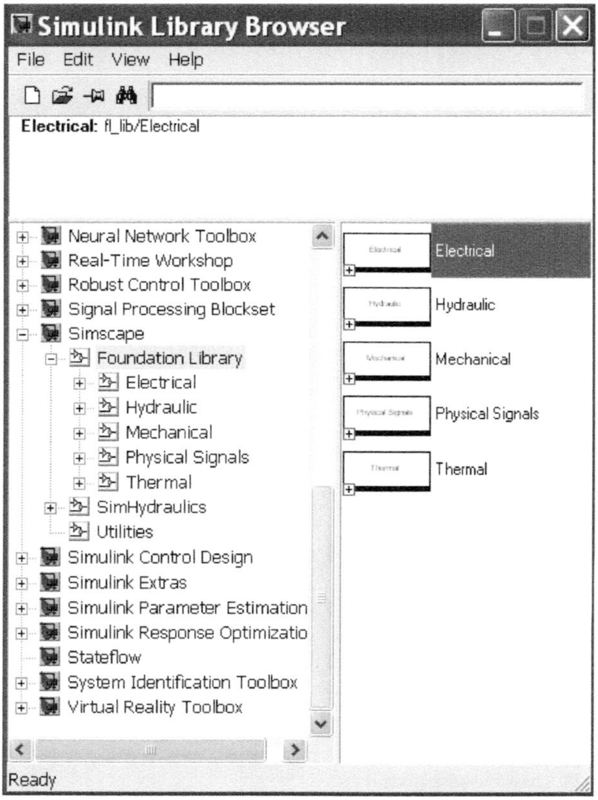

Fig. D.2 The SIMSCAPE Library browser

Once we access SIMSCAPE, a new model must be created by dragging components from the Foundation Library browser to this new model window, by selecting the appropriate library where the searched component is included (fig. D.3).

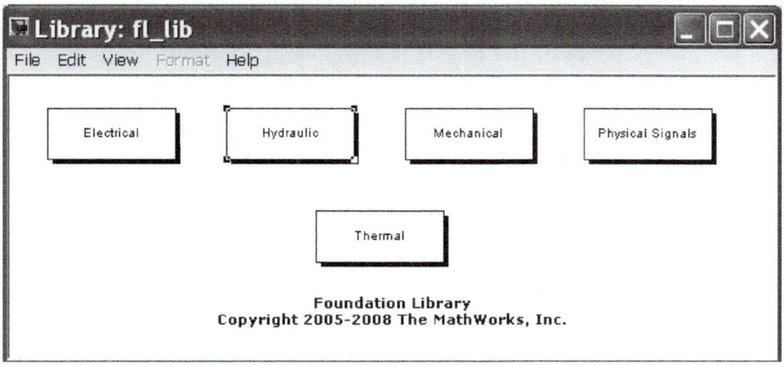

Fig. D.3 The Foundation Library of SIMSCAPE

D.2 Creating a Model

In order to explain the process followed to create a SIMSCAPE model, we have selected an electronic circuit system which performs the PID controller function (fig. D.4).

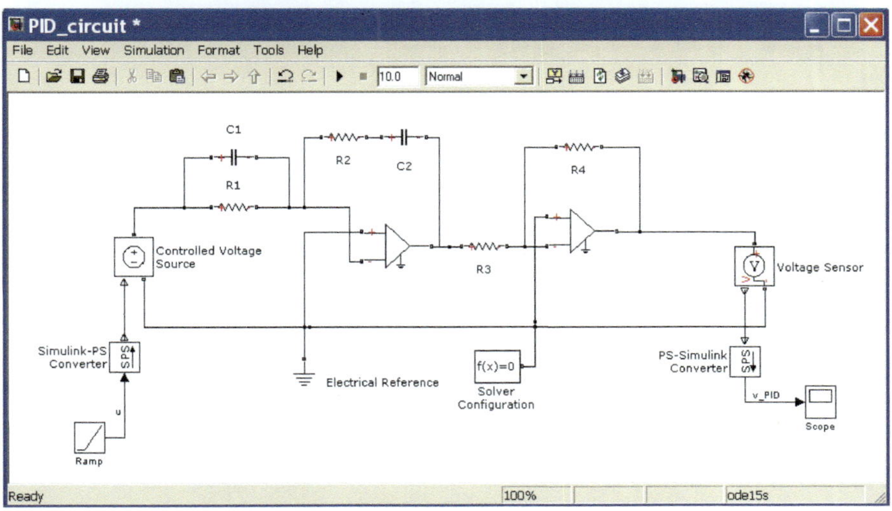

Fig. D.4 PID electronic circuit in SIMSCAPE

In first place we have to collect all components necessary to build the PID controller from the Foundations Library. Afterwards, we will connect the selected components to implement the physical components diagram, also changing the parameters of these components if it is necessary. Finally the system is simulated by selecting the appropriate solver options.

Therefore, as first step, we open a new component model, dragging the constitutive components of the physical components diagram from each one of the components libraries where they are included (fig. D.5).

As second step, we connect these components following the desired connection pattern as indicated in the PID electronic circuit (fig. D.6). Possibly, it will be necessary to flip or rotate some components or to derive some pickoff points.

Afterward, in a third step we have to change the component's parameters according to the values defined in the original PID electronic circuit if it is necessary, adding optionally labels both to identify the signals transferred between components or to name specific components (fig. D.7). Component parameters can also be defined from the MATLAB workspace and used in SIMSCAPE. In this case we have assigned the resistors, capacitors, gains, and ramp slope values as

```
>> R1=153.85*10^3;
>> R2=153.85*10^3;
>> R3=10*10^3;
>> R4=197.1*10^3;
 >> C1=10*10^-6;
>> C2=10*10^-6;
>> ramp_slope=0.1;
```

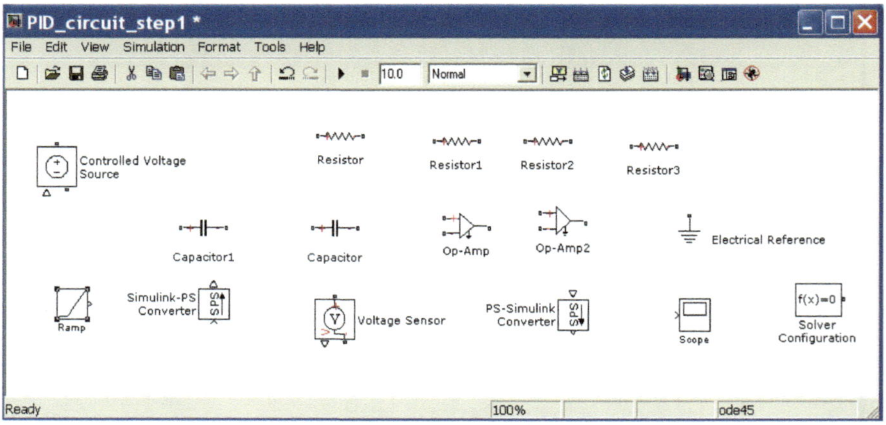

Fig. D.5 Building process for the PID circuit (step 1)

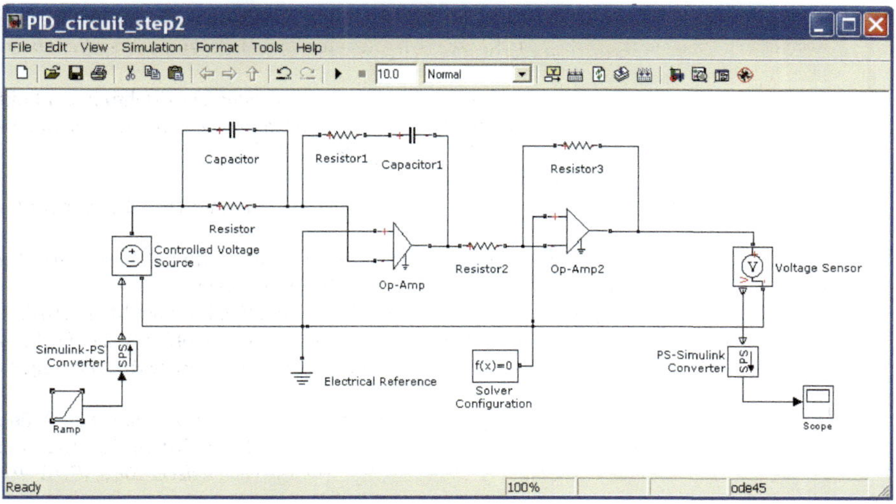

Fig. D.6 Building process for the PID circuit (step 2)

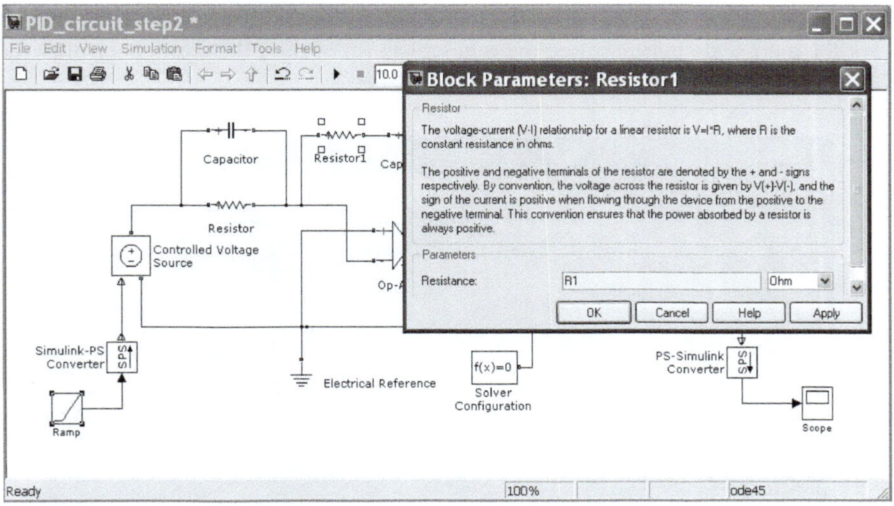

Fig. D.7 Building process for the PID circuit (step 3)

Before simulation is run, we must select the appropriate simulation time and the numerical method used. For this case, we have to choose between *ode15s* and *ode23t*, according to the instructions to be followed by the SIMSCAPE assigned solver (fig. D.8).

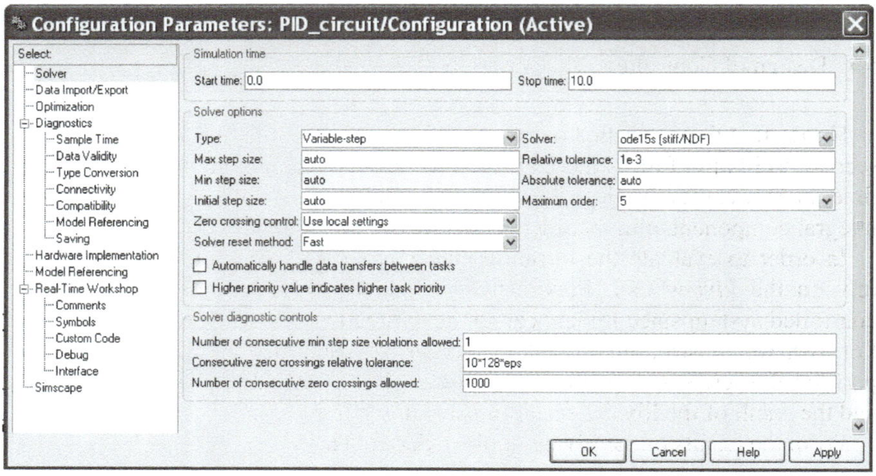

Fig. D.8 Definition of the solver configuration for the PID circuit.

Finally, the simulation is run by double-clicking on the ⊳ icon and output responses can be viewed by double-clicking on the Scope block to view its output. Hit the auto-scale button to see the ramp response of the PID circuit controller for a ramp slope of 0.1 (fig. D.9).

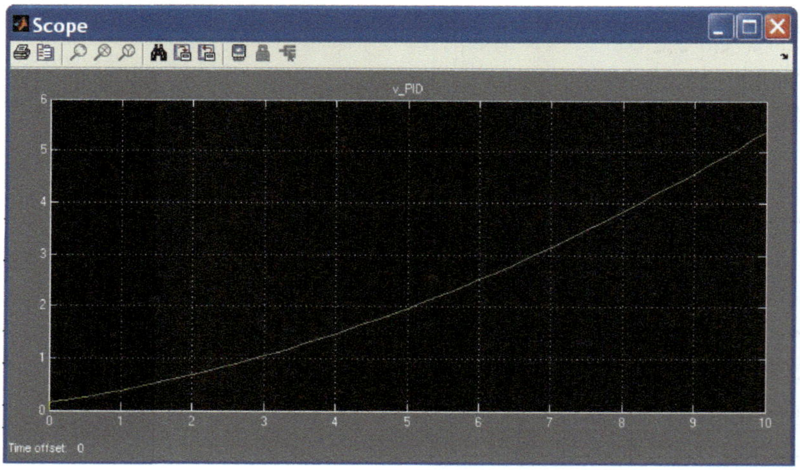

Fig. D.9 Ramp response of the PID using SIMSCAPE

D.3 SIMSCAPE Libraries

SIMSCAPE contains a large number of components from which physical models can be built. These components are arranged in Components Libraries which are accessed by the main SIMULINK window.

The set of Component Libraries we are going to describe contains the main blocks used for analysis, modeling and control of dynamic systems, mainly composed of electrical, hydraulic, mechanical, and thermal components, together with physical signal blocks utilized to define components with more complicated physical relations.

D.3.1 Electrical Library

This section is an introduction to the Electrical Library of SIMSCAPE. We will illustrate the components included in this library (fig. D.10).

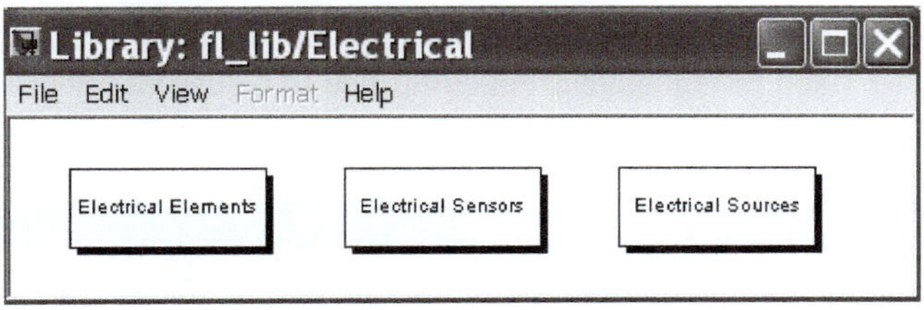

Fig. D.10 The SIMSCAPE Electrical Library

The Electrical Library is composed of three sub-libraries, termed the Electrical Elements, Electrical Sensors, and Electrical Sources libraries.

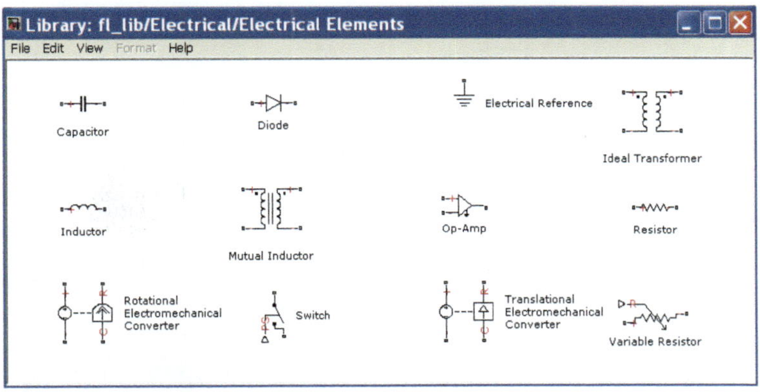

Fig. D.11 The SIMSCAPE Electrical Elements sub-library

The Electrical Elements library contains some of the common elements which constitute the electric and electronic networks, both linear (capacitor, inductance, resistance, etc.) and nonlinear (diode, switch, etc.) together with the electromechanical converters (fig. D.11).

The Electrical Sensors library contains both a voltage and a current sensor to be placed in parallel or series with the element whose magnitude is being measured (fig. D.12).

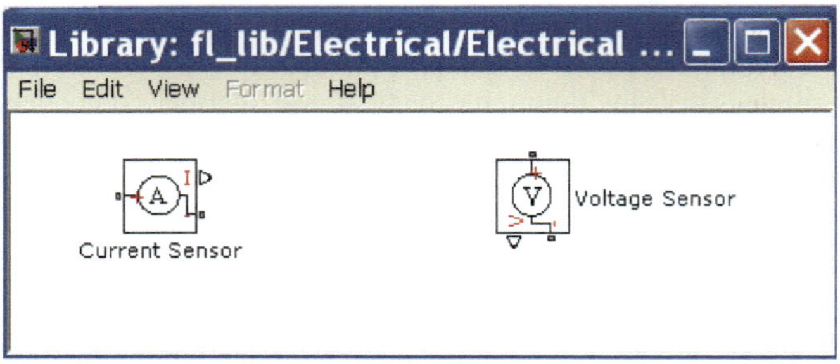

Fig. D.12 The SIMSCAPE Electrical Sensors sub-library

The Electrical Source library contains both voltage and current (AC and DC) sources, either fixed or voltage/current controlled to be used as electric sources for the model (fig. D.13).

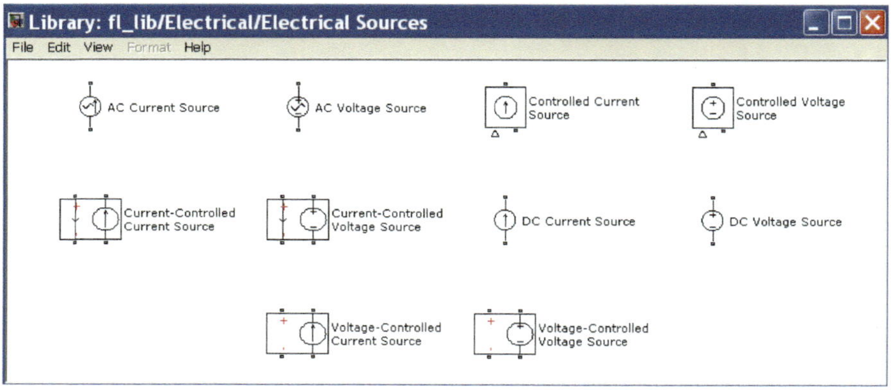

Fig. D.13 The SIMSCAPE Electrical Sources sub-library

D.3.2 Hydraulic Library

This section is an introduction to the Hydraulic Library of SIMSCAPE. We will illustrate the components included in this library (fig. D.14).

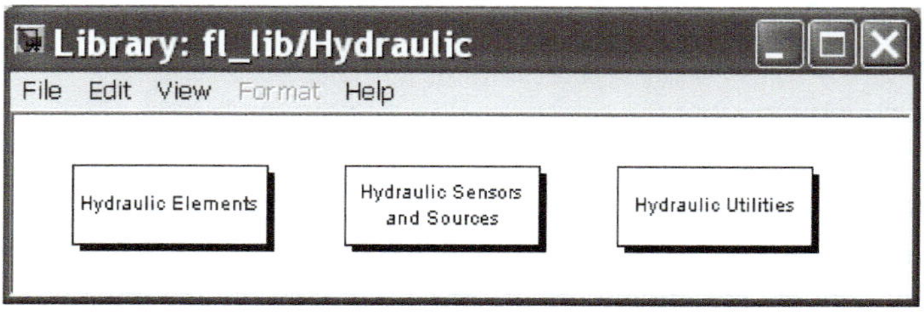

Fig. D.14 The SIMSCAPE Hydraulic Library

The Hydraulic Library is composed by three sub-libraries, termed the Hydraulic Elements, Hydraulic Sensors and Actuators, and Hydraulic Utilities libraries.

The Hydraulic Elements library contains some of the common elements which constitute the hydraulic circuits (resistance, inertance, capacitance,..) together with the hydromechanical converters (pump and piston) (fig. D.15).

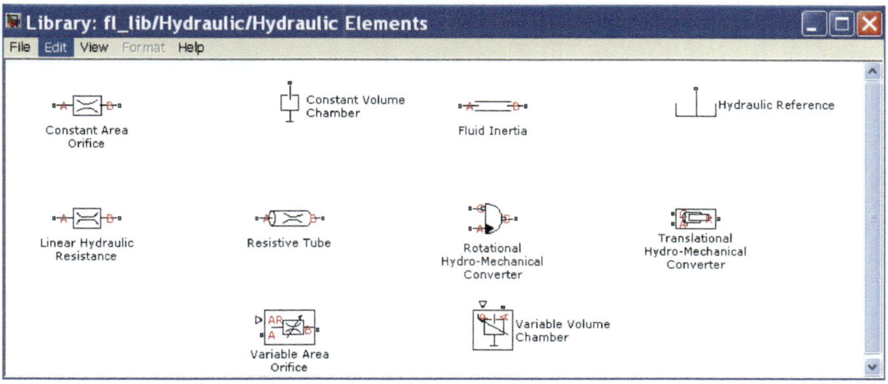

Fig. D.15 The SIMSCAPE Hydraulic Elements sub-library

The Hydraulic Sensors and Actuators library contains both pressure and flow sensors and sources in order to measure or actuate over the hydraulic circuit elements (fig. D.16).

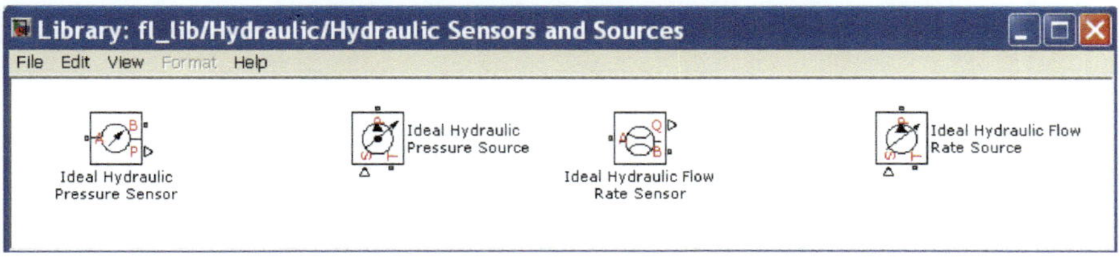

Fig. D.16 The SIMSCAPE Hydraulic Sensors and Actuators sub-library

Finally, a element to define the characteristics of fluid used by the hydraulic circuit is also included, into the Hydraulic Uilities library (fig. D.17)

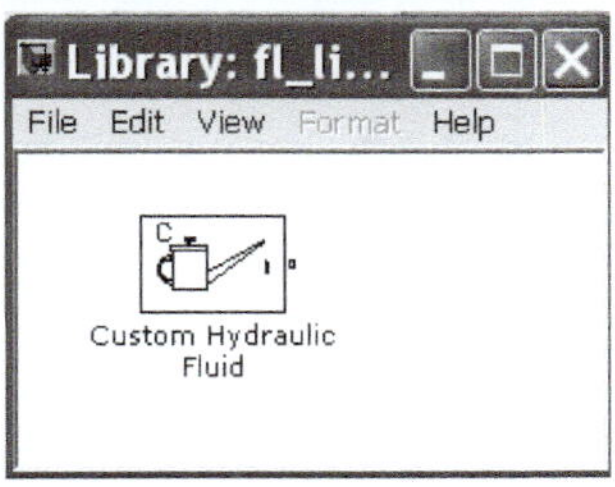

Fig. D.17 The SIMSCAPE Hydraulic Utilities sub-library

D.3.3 Mechanical Library

This section is an introduction to the Mechanical Library of SIMSCAPE. We will illustrate the components included in this library (fig. D.18).

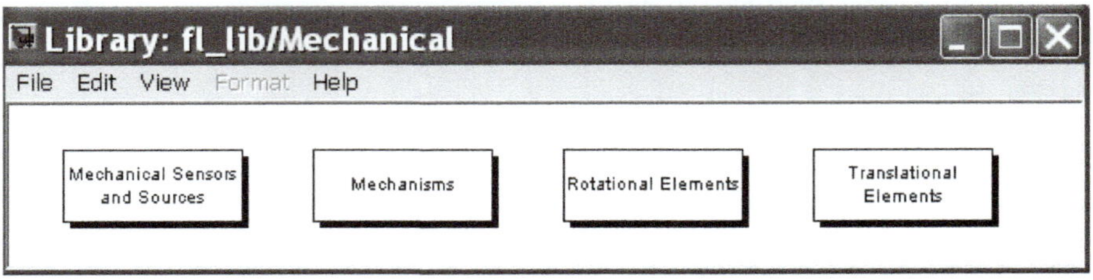

Fig. D.18 The SIMSCAPE Mechanical Library

The Mechanical Library is composed by four sub-libraries, termed the Translational Elements, Rotational Elements, Mechanical Sensors and Sources, and Mechanism libraries.

The Translational Elements library contains some of the common elements which constitute the translational mechanical systems (mass, spring, damper,…) (fig. D.19), while the Rotational Elements library contains some of the common elements which constitute the rotational mechanical systems (inertia, spring, damper,…) (fig. D.20).

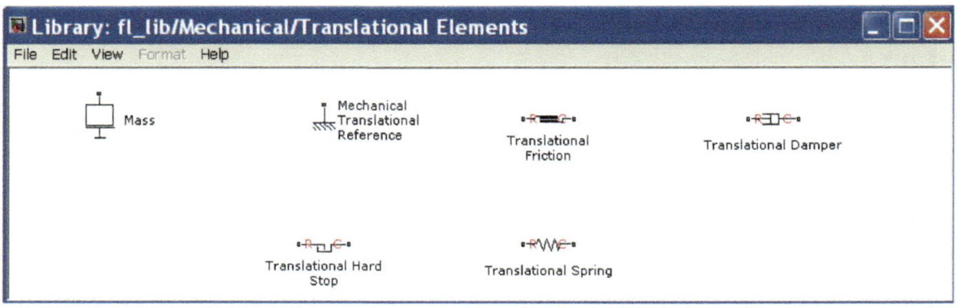

Fig. D.19 The SIMSCAPE Mechanical Translational Elements sub-library

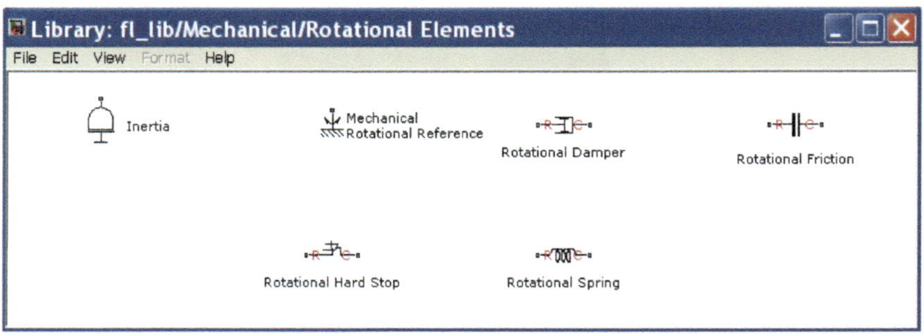

Fig. D.20 The SIMSCAPE Mechanical Translational Elements sub-library

The Mechanical Sensors and Sources library contains both translational and rotational position/velocity sensors as well as force/torque sensors. Besides, contains force/torque sources in order to actuate over the mechanical circuit elements, as well as velocity sources (fig. D.21).

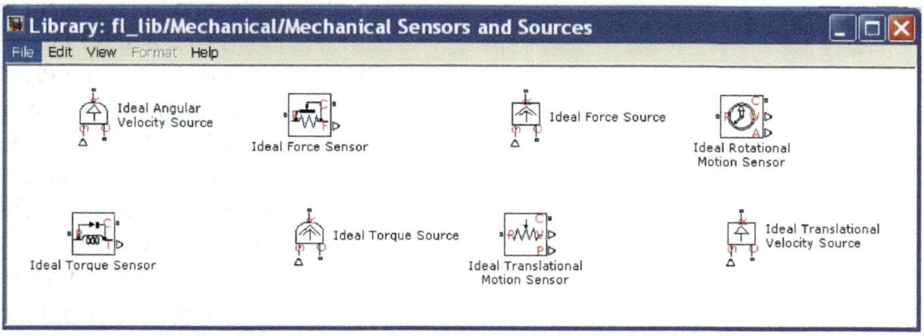

Fig. D.21 The SIMSCAPE Mechanical Sensors and Sources sub-library

Additionally, a Mechanism library where mechanical transformers (gear and lever) and ideal pulleys can be found, is included (fig. D.22).

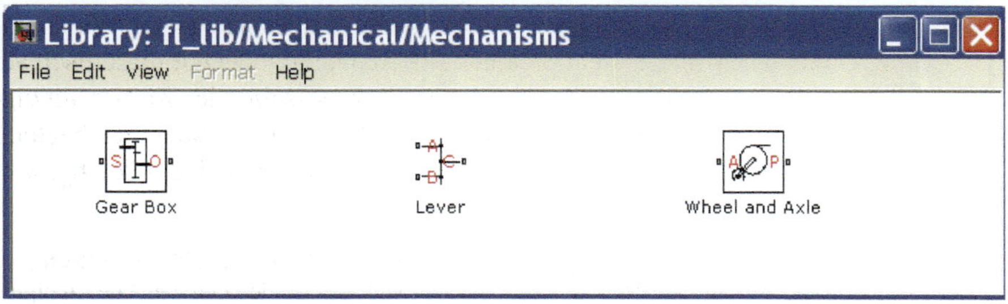

Fig. D.22 The SIMSCAPE Mechanisms sub-library

D.3.4 Thermal Library

This section is an introduction to the Thermal Library of SIMSCAPE. We will illustrate the components included in this library (fig. D.23).

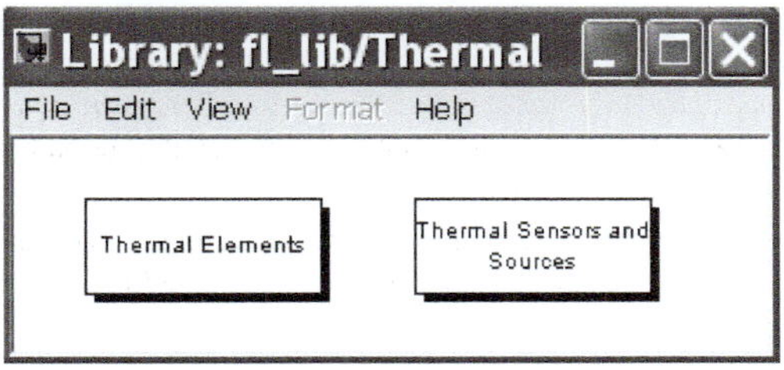

Fig. D.23 The SIMSCAPE Thermal Library

The Thermal Library is composed of two sub-libraries, termed the Thermal Elements, and Thermal Sensors and Sources libraries.

The Thermal Elements library contains some of the common elements which constitute the thermal systems (mass and heat trasnfer mechanism) (fig. D.24).

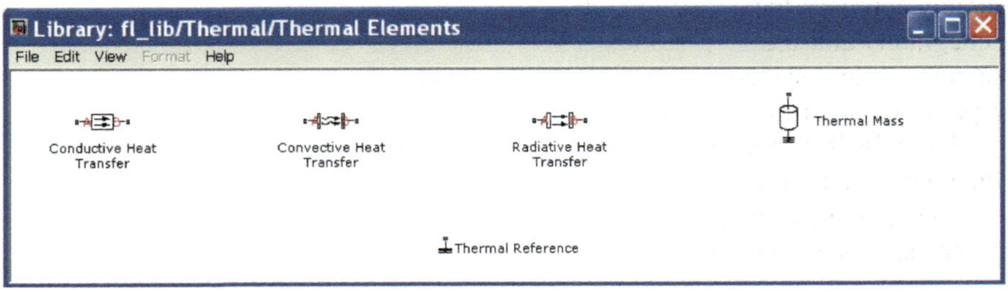

Fig. D.24 The SIMSCAPE Thermal Elements sub-library

The Thermal Sensors and Sources library contains both tempearture and heat flow sensors and sources in order to measure or actuate over the thermal circuit elements (fig. D.25).

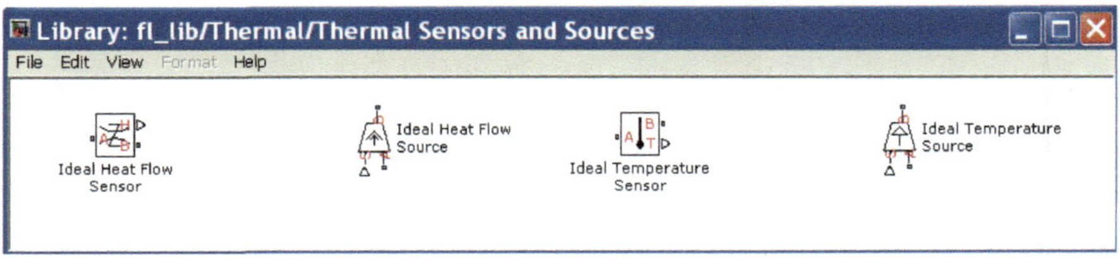

Fig. D.25 The SIMSCAPE Thermal Sensors and Sources sub-library

D.3.5 Physical Signal Library

This section is an introduction to the Physical Signal Library of SIMSCAPE, which is utilized to create equations, by using linear and nonlinear operators, functions, lookup tables, etc. These equations are solved simultaneously with physical system, and unit conversions are handled automatically.

We will illustrate the blocks included in this library and its function (fig. D.26). These blocks are similar to those described in the SIMULINK Libraries, and perform the same function, by using physical signal instead of dimensionless variables as SIMULINK. In this way, it includes linear blocks (adder, gain, integrator,…) and nonlinear also (divider, math function, table lookup, saturation,…) as it is shown in fig. D.27.

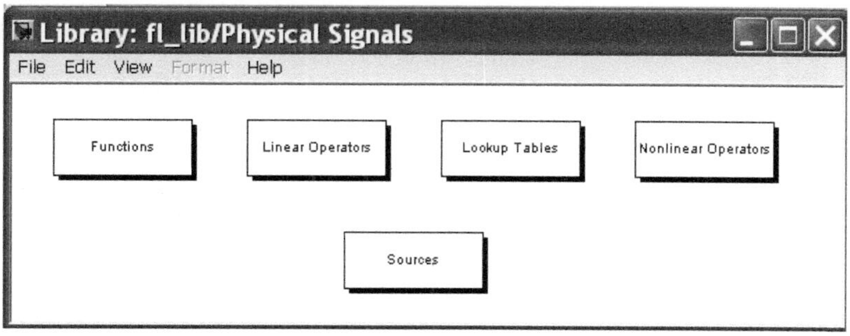

Fig. D.26 The SIMSCAPE Physical Signals Library

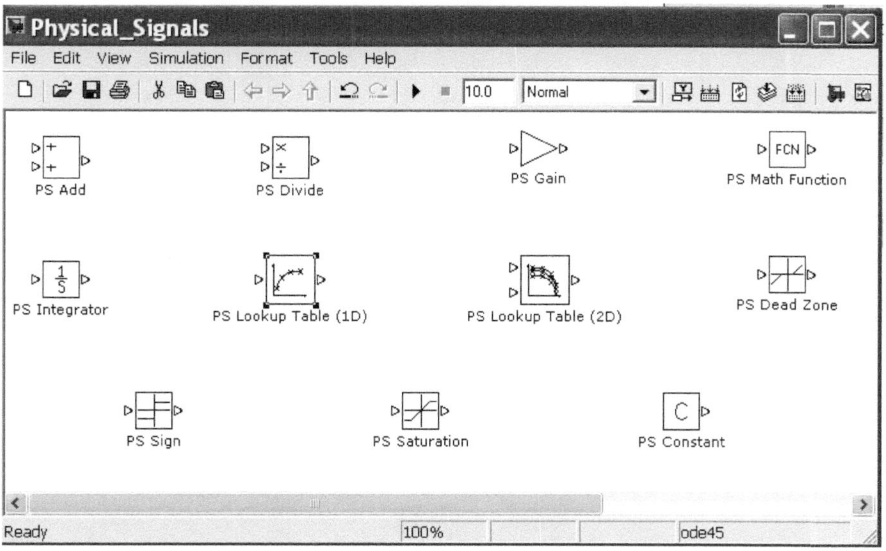

Fig. D.27 Blocks included into the SIMSCAPE Physical Signals Library

D.3.6 Utilities Library

This section is an introduction to the Utilities Library of SIMSCAPE, which contains general purpose blocks as the PSS and SPS converters from SIMSCAPE to SIMULINK or vice versa respectively, together with the input/output ports to define subsystems in SIMSCAPE. Additionally, the solver block, which is necessary to be attached to the physical components diagram is included, thus the solver can be applied with the numerical methods to integrate the dynamic system equations (fig. D.28).

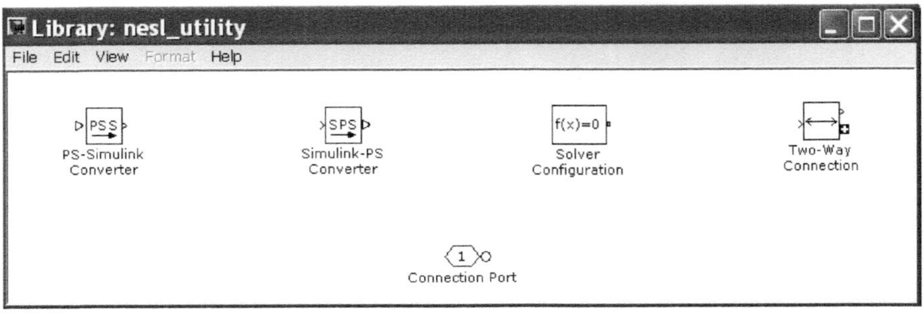

Fig. D.28 The SIMSCAPE Utilities Library

Index

R

S

GPSR Compliance

The European Union's (EU) General Product Safety Regulation (GPSR) is a set of rules that requires consumer products to be safe and our obligations to ensure this.

If you have any concerns about our products, you can contact us on ProductSafety@springernature.com

In case Publisher is established outside the EU, the EU authorized representative is:

Springer Nature Customer Service Center GmbH
Europaplatz 3
69115 Heidelberg, Germany

Batch number: 10614940

Printed by Printforce, the Netherlands